## ALICE IN NUMBEl

# Alice in Numberland
## A Students' Guide to the Enjoyment of Higher Mathematics

John Baylis
Rod Haggarty

**M**
MACMILLAN

© John Baylis and Rod Haggarty 1988

All rights reserved. No reproduction, copy or transmission of
this publication may be made without written permission.

No paragraph of this publication may be reproduced, copied or
transmitted save with written permission or in accordance with
the provisions of the Copyright, Designs and Patents Act 1988,
or under the terms of any licence permitting limited copying
issued by the Copyright Licensing Agency, 90 Tottenham Court
Road, London W1P 9HE.

Any person who does any unauthorised act in relation to this
publication may be liable to criminal prosecution and civil
claims for damages.

First published 1988 by
MACMILLAN EDUCATION LTD
Houndmills, Basingstoke, Hampshire RG21 2XS
and London
Companies and representatives
throughout the world

ISBN 0–333–44242–3

A catalogue record for this book is available
from the British Library

Printed in Hong Kong

Reprinted 1991

To Daniel, Gwen, Siân and Sarah

# CONTENTS

*Introduction* vii

1. Alice in Logiland—in which we meet Alice, Tweedledee and Tweedledum, and Logic — 1

2. Unique factorisation—in which trivial arithmetic reveals a glimpse of hidden depths — 14

3. Numbers—in which we abandon logic to achieve understanding, then use logic to deepen understanding — 21

4. The real numbers—in which we find holes in the number line and pay the price for repairs — 36

5. A variety of versions and uses of induction—in which another triviality plays the lead — 52

6. Permutations—in which ALICE is transformed — 72

7. Nests—in which the rationals give birth to the reals and the scene is set for arithmetic in $\mathbb{R}$ — 89

8. Axioms for $\mathbb{R}$—in which we invent Arithmetic, Order our numbers and Complete our description of the reals — 98

9. Some infinite surprises—in which some wild sets are tamed, and some nearly escape — 111

| | | |
|---|---|---|
| **10** | Sequences and series—in which we discover very odd behaviour in even the smallest infinite set | 141 |
| **11** | Graphs and continuity—in which we arrange a marriage between Intuition and Rigour | 161 |

*Suggestions for further reading*   177

*Hints and solutions to selected exercises*   180

*Index*   203

# INTRODUCTION

## Failure to read this can damage your understanding

> 'It saddens me that educated people don't even know that my subject exists.'
>
> Paul R. Halmos

What is our purpose in writing this book? Quite simply, to help students of mathematics *enjoy* their work. It seems to us that some students are good at mathematics and just get on with it, others tolerate it for its promised usefulness, but few actively enjoy it except in the rather trivial sense of preferring success to frustration.

We have set out quite deliberately to write an entertaining book whose content is serious mathematics. The professional mathematician will be familiar with the idea that entertainment and serious intent are not incompatible: the problem for us is to ensure that our readers will enjoy the entertainment but not miss the mathematical point, which is why we suggest that you regard this introduction as compulsory reading! None of the light-hearted sections are there just for entertainment; they are all meant to illuminate important mathematics. Now that we have alerted you to this, perhaps you will forgive us if we occasionally 'explain the point of the joke'. If you don't need the explanation, so much the better.

What we do assume of our readers is that they already have some interest in and commitment to mathematics as either students or teachers, and that they have some feeling that its pursuit could be enjoyable. Our domain is pure mathematics at the general level of sixth-form and early-undergraduate work, and the path we intend to take through this domain is deliberately not direct. It is winding, it branches at many points and it is often self-intersecting. It would be possible to plan such a route in which the general scenery and specific landmarks are the fascinating applications of the subject, but we have chosen a different route based on showing mathematics as interesting in its own right. Actually, 'interesting' is rather too bland a word to express the feelings we hope this book will foster; 'excitement' and 'exhilaration' would be more accurate. People find it hard to believe that mathematicians can feel the same sort of emotions and satisfactions for their subject as those with poetic, musical or even sporting inclinations can feel for theirs. Nevertheless it is true, and this has been a factor to consider in planning the route.

Another major thread running throughout the book is the idea of proof.

# INTRODUCTION

It is fair to say that the criteria for and methods of proof make mathematics unique among all academic disciplines—even the physical sciences. Sadly, proof is an aspect of mathematical education which is often neglected, and the effect of this neglect can be severe. It means that we sell a subject to prospective students under false pretences, and when they find that university mathematics is very largely about proof and that they don't like it, they have no Trade Descriptions Act to protect them from a fairly miserable three years. There may be a complementary, equally sad effect—that of some potential students never embarking upon a mathematics degree precisely because no one ever demonstrated to them that the search for proof was a pleasure and not a chore of mathematical life, though, for obvious reasons, evidence for this is harder to obtain! To appreciate proof, students must be properly prepared, at sixth-form level and even before—otherwise a subject such as first-year analysis will appear to be all about proving obvious, not very exciting facts by perversely difficult methods. What, then, is an obvious fact, and what do we gain by proving it? Chapter 2 goes some way towards providing an answer.

At the other extreme, there are some unbelievable (yet true) statements about such basics as numbers and graphs, and we plan to highlight these too. Profundity hidden in the obvious and shock and wonder generated by the totally unexpected provide the satisfactions and excitement of mathematics, and it is natural that those who have experienced them should feel some missionary zeal in spreading the word to others, especially as so many have a view of mathematics as 'sums, only harder'. Some who are 'good at mathematics' sometimes try to tell others that the subject is easy. Such people are not mathematicians; they are those who are satisfied with solving problems they find easy anyway! Real satisfaction, whether in marathon running, mountaineering, mathematics or music, comes from accepting the challenge of hard problems, with some exercise of judgement to avoid selecting too many unrealistically difficult ones!

But in mathematics, solving problems is not sufficient. If every problem required the invention of a totally new technique, it would become impossibly difficult rather than just difficult, and satisfaction would rapidly diminish as the feeling that the subject had no intrinsic unity increased. For this reason unity is another important theme. To take account of it, a decision had to be made about exactly which proof techniques to include, and the answer had to be those proofs based on principles which, although very simple, have profound consequences in a variety of different parts of mathematics. We have, in general, avoided one-off tricks except where they are so neat that they cry out for inclusion.

Finally, although the book has, we hope, what computing colleagues would call a reader-friendly style, this is by no means incompatible with mathematical rigour, and we intend to *do* a lot of serious mathematics as well as talk about it! Some of the questions we ask (such as 'What is a real number?' or 'What counts as a proof?') turn out to be profound, and the hard work they entail is shared between deciding what the significance of the question is and what it would mean to answer it, and doing the detailed working through to achieve the answer. At various points there are gaps indicated thus [**m.n**] to be filled by the reader, and hints are provided in the appendix. Ends of proofs are signalled by the Halmos tombstone symbol, □, and particularly important results and definitions are enclosed by a box.

The book does not have to be read in the order in which it is written. Depending on the knowledge and interests of the reader, there are convenient places to change the route, and these, too, are signposted. It is important to point out that this is not a systematic textbook on any particular branch of mathematics, but our hope is that its overall effect will be to provide motivation to tackle more formal texts, ability to understand and enjoy them, and a healthy critical attitude which will demand close attention both to rigour and to a sound intuitive base.

'The purpose of rigour is to legitimate the conquests of intuition' (Jacques Hadamand, 1865–1962)—and to reveal its not-too-infrequent blunders, we should perhaps add!

We extend our thanks to Macmillan Education's former mathematics editor, Peter Oates, for his help and patience; to Cathy Crewe, Angela Fullerton and Diane Holloway for transforming illegible scrawl; and to each other for mutual support. As for remaining errors, I blame him and he blames me.

*Nottingham and Abingdon*, 1987
J.B.
R.H.

# 1 Alice in Logiland—in which we meet Alice, Tweedledee and Tweedledum, and Logic ...

> 'Contrariwise', continued Tweedledee, 'if it was so, it might be; and if it were so, it would be; but as it isn't, it ain't. That's logic.'
> Lewis Carroll

## 1.1 LOGIC

In *Alice through the Looking-glass* Alice meets Tweedledee and Tweedledum in the forest. The brothers look so much alike that Alice cannot tell them apart. In our version of the story one of the brothers tells lies on Mondays, Tuesdays and Wednesdays but tells the truth on the other days of the week. The other brother lies on Thursdays, Fridays and Saturdays but is truthful on the remaining days of the week. One day Alice meets the two in the forest, and the following conversation takes place:

FIRST BROTHER: I am Tweedledum.
SECOND BROTHER: I am Tweedledee.
ALICE: Ah, so it's Sunday.

Alice is correct, of course, since, if one of the brothers is lying, the other is telling the truth and so they would both claim to be Tweedle—. Hence, both brothers are telling the truth and this only happens on Sundays. Our analysis depends on the fact that the sentences uttered by the brothers are either true or false. Such a sentence is called a *statement*. In logic we shall only allow statements, and so phrases such as 'I am lying' or 'What time is it?' will not count as statements; the latter phrase since it is a question, but why not the former? [**1.1**] The conversation continues:

FIRST BROTHER: I am Tweedledee or it is raining.
ALICE: Oh dear, I've forgotten my umbrella.

Now, remember it's still Sunday, so both brothers are telling the truth. Hence, the first brother is Tweedledum (from the earlier conversation). So it must be raining, since the statement 'I am Tweedledee or it is raining' is true and the first brother is not Tweedledee. Tweedledum's statement is an example of a *compound statement* being two simple statements joined by the *connective* 'or'. Other connectives used for building up compound statements are 'and' and 'not'. We give these three connectives precise logical meanings before continuing our adventures with Alice.

First, a statement which is true will be given the truth value T and one which is false will be labelled F.

Given a statement $P$, the statement that $P$ does not hold, abbreviated $\sim P$, will have the opposite truth value to $P$'s. This definition can be neatly displayed in a truth table (see Figure 1.1).

| $P$ | $\sim P$ |
|---|---|
| T | F |
| F | T |

**Fig. 1.1**

Given another statement $Q$, the compound statement '$P$ or $Q$', written as $P \vee Q$, will be true provided that at least one of $P$ or $Q$ is true. Notice that 'or' is being used in the inclusive sense of 'and/or' and not in the exclusive sense of 'one but not both'. The compound statement '$P$ and $Q$', written as $P \wedge Q$, will be true only when both $P$ and $Q$ are true statements. Again this information is easily displayed in truth tables (see Figure 1.2).

| $P$ | $Q$ | $P \vee Q$ | $P$ | $Q$ | $P \wedge Q$ |
|---|---|---|---|---|---|
| T | T | T | T | T | T |
| T | F | T | T | F | F |
| F | T | T | F | T | F |
| F | F | F | F | F | F |

**Fig. 1.2**

Meanwhile, back in the forest, on some day the following week, Alice meets the brothers again. Alice has had so many adventures that she has forgotten the day of the week.

FIRST BROTHER: I am Tweedledum and he is Tweedledee.
ALICE: Then I'm sure I don't know who you are.
SECOND BROTHER: I am not Tweedledee.
ALICE: As if that helps.

Alice, being rather good at logic, is of course correct. We cannot decide which is which from the last conversation. [**1.2**]

FIRST BROTHER: Contrariwise, I am not Tweedledum or he is not Tweedledee.
ALICE: Haven't you just said that?

Perhaps the brothers have at last confused Alice! Can we decide whether or not the first brother has said the same thing in a different way? Of course 'contrariwise' means 'it is not the case that', and so we are dealing with the negation of some statement. To analyse this problem, we proceed as follows.

Let $P$ stand for the statement 'I am Tweedledum' and $Q$ abbreviate the statement 'He is Tweedledee'. The two statements made by the first brother are thus $P \wedge Q$ in the first instance, and then $\sim(\sim P \vee \sim Q)$. These two compound statements will be logically the same if they have the same truth values, no matter what the truth values of the constituent statements. This can be confirmed, or otherwise, by constructing their truth tables (see Figure 1.3). Thus $P \wedge Q$ is *logically equivalent* to $\sim(\sim P \vee \sim Q)$.

| $P$ | $Q$ | $\sim P$ | $\sim Q$ | $\sim P \vee \sim Q$ | $\sim(\sim P \vee \sim Q)$ | $P \wedge Q$ |
|---|---|---|---|---|---|---|
| T | T | F | F | F | T | T |
| T | F | F | T | T | F | F |
| F | T | T | F | T | F | F |
| F | F | T | T | T | F | F |

**Fig. 1.3**

Later in the week, Alice meets her friends again.

FIRST BROTHER: If I'm Tweedledum, then he's Tweedledee.
SECOND BROTHER: If he's Tweedledee, then I'm Tweedledum.
ALICE: Oh, so it's Sunday again and still I don't know which of you is which.

Once again Alice's logic is faultless. The first brother has said 'If $P$ then $Q$', usually written as $P \Rightarrow Q$. Now this statement is only false when the conclusion $Q$ is false but the hypothesis $P$ is true. This is because we cannot deduce, in a logically sound manner, a false result from a true one. This information appears in the truth table for the connective '$\Rightarrow$' (see Figure 1.4).

It is worth while leaving our friends for a while, since there is something

| P | Q | P⇒Q |
|---|---|-----|
| T | T | T |
| T | F | F |
| F | T | T |
| F | F | T |

Fig. 1.4

distinctly odd about the truth table in Figure 1.4. To see this, consider the following three statements, which, according to our truth table, are deemed to be true:

(a) If $2+2=4$, then grass is usually green.
(b) If $2+2=5$, then grass is usually green.
(c) If $2+2=5$, then *Alice through the Looking-glass* was written by Queen Victoria.

To call such statements true is clearly nonsense, and so we must admit that the language of everyday conversation and formal (mathematical) logic are not the same thing. We have been guilty of suggesting that they are—indeed, guilty of implying that $\sim$, $\vee$, $\wedge$ and $\Rightarrow$ are just abbreviations for the normal conversational phrases 'not', 'or', 'and' and 'implies'. This is not the case, because statements of the form $P \vee Q$ and $P \Rightarrow Q$ are reasonable utterances in normal conversation only if there is some connection in context and meaning between $P$ and $Q$. Alice only gets away with remarks such as 'It's Sunday and I still don't know which of you is which' because she (and we) know that there is some connection between the day of the week and the truthfulness of the brothers. Even Alice could not avoid ridicule if she habitually made pronouncements like examples (a)–(c) above. In formal logic no such connection is assumed. Everything about a statement is abstracted away, except its truth or falsity. Even so, the truth of $P \Rightarrow Q$ when $P$ is false, whatever $Q$ is, is a little difficult to swallow. Consider the statement 'If it is Tuesday I drink wine', and ask under what circumstances this could be false. It is shown to be false only by my not drinking wine on a Tuesday. In other words, the original statement is intentionally equivalent to 'It is not the case that I do not drink wine on Tuesday'. This is always the way in which mathematical implication works. That is, $P \Rightarrow Q$ is always taken to mean $\sim (P \wedge \sim Q)$. Verify the logical equivalence of these last two logical expressions. [**1.3**] As a concrete example see Figure 1.7, where $\triangle ABC$ is right-angled at A with side lengths $x, y$ and $z$, as indicated. Pythagoras' theorem states: 'If $\triangle ABC$ is right-angled, then $x^2 + y^2 = z^2$'. Equivalently, 'No triangle can be right-angled and have $x^2 + y^2 \neq z^2$'. To take a more extreme example, consider the following $P \Rightarrow Q$ statement: 'If $n$ is a whole number bigger than 69 and less than 70, then $n$ is a prime number'. The judgement of mathematical logic is that this theorem is a true one—true by default if you like, because $P$ is never true, so the statement is a totally vacuous one.

Returning to Alice and her last conversation with the two brothers. The first brother's statement $P \Rightarrow Q$ is true; otherwise both brothers are Tweedledum! But from Figure 1.4 this tells us nothing about either the truth of $P$ or the truth of $Q$. Similarly, the second brother's statement $Q \Rightarrow P$ is also true. Alice is correct; it is Sunday but we still cannot identify the brothers.

FIRST BROTHER: If he's not Tweedledee, then I'm not Tweedledum.
ALICE: Oh, there you go again, repeating yourself.

Show that Alice is again correct. In other words, show that $(\sim Q \Rightarrow \sim P)$, the *contrapositive* of $P \Rightarrow Q$, is logically equivalent to $P \Rightarrow Q$. [**1.4**] The statement $Q \Rightarrow P$ is called the *converse* of the statement $P \Rightarrow Q$ and the compound statement $(P \Rightarrow Q) \wedge (Q \Rightarrow P)$ is written as $P \Leftrightarrow Q$. This is often read as '$P$ if and only if $Q$'. Its truth table can easily be deduced from the truth tables for 'implies' and 'and' (see Figure 1.5). We can see immediately that $P \Leftrightarrow Q$ is true whenever $P$ and $Q$ are logically equivalent statements.

| $P$ | $Q$ | $P \Leftrightarrow Q$ |
|---|---|---|
| T | T | T |
| T | F | F |
| F | T | F |
| F | F | T |

Fig. 1.5

## 1.2 THEOREMS

Many mathematical theorems consist of a collection of hypotheses and a conclusion. They are thus of the logical form 'if $P$ then $Q$'. The proof of such theorems consists of a demonstration that $P \Rightarrow Q$ is a true statement. There are three common methods of proof:

(1) Assume that $P$ is true and deduce by some process that $Q$ is true.

(2) Assume that $Q$ is false and deduce that $P$ is false. In other words, prove the contrapositive.

(3) Assume that $P$ is true and that $Q$ is false, and deduce some obviously false statement. In other words, show that $\sim (P \wedge \sim Q)$ is true. This is logically equivalent to $P \Rightarrow Q$ is true, as you showed earlier.

Proofs of type (3) are surprisingly common and the derivation of some obvious contradiction is called *reductio ad absurdum*. We shall see many examples in the sequel. Some theorems have the form $P \Leftrightarrow Q$, which is really two theorems in one. In this case we have to prove that $Q \Rightarrow P$ holds as well as $P \Rightarrow Q$. As an example of the above methods of proving $P \Rightarrow Q$, we prove the following result concerning positive whole numbers (called *natural numbers*).

**Theorem 1.1** If $n^2$ is odd, then $n$ is odd.

We let $P$ denote '$n^2$ is odd' and $Q$ denote '$n$ is odd'.

*Proof 1* By the uniqueness of prime factorisation (UPF: see Chapter 2), $n = p_1 p_2 \ldots p_r$ for primes $p_1, p_2, \ldots, p_r$ (not necessarily all distinct). Hence, $n^2 = p_1^2 p_2^2 \ldots p_r^2$. But $n^2$ is odd and so none of the $p_i$'s can equal 2. Thus, $n$ is clearly odd. We have demonstrated directly that $P \Rightarrow Q$ is true. □

*Proof 2* If $n$ is an even number, then $n = 2m$ for some other whole number $m$. Now $n^2 = 2(2m^2)$ is also even. This establishes $P \Rightarrow Q$ by proving the contrapositive $\sim Q \Rightarrow \sim P$. □

*Proof 3* If $n^2$ is odd and $n$ is even, then $n^2 + n$ is odd. Hence, $n(n+1)$ is odd. But $n$ and $n+1$ are consecutive whole numbers and so one of them must be even. In other words, $n(n+1)$ is also even. This is clearly impossible. We have derived a contradiction from the assumption that $P \wedge \sim Q$ is true. Hence, $\sim (P \wedge \sim Q)$ is true. This establishes the truth of $P \Rightarrow Q$. □

Care must be exercised when constructing proofs by contradiction. For example, what is wrong with the following?[**1.5**]

**Theorem 1.2** 1 is the largest natural number.
*Proof* Suppose that 1 is not the largest natural number. Let $n$ be the largest natural number. Now $n + 1$ is a larger natural number, which is a contradiction (or is it?) □

Mathematical sentences often involve a variable $x$. If such a sentence is either true or false for a fixed value of $x$, we call the sentence a *predicate* and denote it by $P(x)$. Thus, for any particular value of the variable $x$, $P(x)$ yields a statement. For example, consider the sentence $P(x)$ given by '$x$ is a real number such that $2x = x^2$'. From elementary algebra we can show that $P(0)$ is true; $P(2)$ is true but $P(x)$ is false for all other values of $x$. Predicates can be converted into statements involving variables by prefixing them with *quantifiers*. There is an *existential quantifier*, 'there exists', abbreviated by ∃, and the *universal quantifier*, 'for all', abbreviated by ∀. Theorems of the form '$\forall x, P(x)$' appear very daunting to prove, since we have to demonstrate that $P(x)$ is true for every $x$ in some (possibly infinite) collection of objects. On the other hand, theorems of the form '$\exists x, P(x)$' appear somewhat less demanding in that we (only) have to establish the truth of $P(x)$ for one single value of $x$. Many of the most beautiful theorems of mathematics are existential in character. One could argue that such theorems are the most important. For example, what is the point in trying to solve some algebraic or differential equation when the existence of even one solution is not guaranteed? We mentioned above that the proofs of 'for all' type and 'there exists' type theorems were apparently different in nature. In practice they are not, for suppose we have formulated a statement of the form '$\forall x, P(x)$' but later on we doubt its veracity. We now seek to disprove our own conjecture. (Not an uncommon situation for working mathematicians!) We need to find one value of $x$ for which $P(x)$ is false. The $x$ in question—and there may be more than one possible—is called a *counterexample* to our original claim. Counterexamples play a major role in mathematical research. To summarise, then, the falsity of '$\forall x, P(x)$' is established by demonstrating the truth of '$\exists x, \sim P(x)$'. So the *contradictory* of a universal statement is an associated existential statement, and conversely, as in Figure 1.6.

| Statement | Contradictory |
|---|---|
| $\forall x, P(x)$ | $\exists x, \sim P(x)$ |
| $\exists x, P(x)$ | $\forall x, \sim P(x)$ |

Fig. 1.6

We consider three questions posed by the famous French mathematician Pierre de Fermat (1601–1665). Mathematicians often suspect that certain statements are true, on the basis of experimental evidence and past experience, while having no idea how to prove them. Such statements are called *conjectures* until such time as they are resolved. Fermat was the son of a leather merchant and he was trained as a lawyer. He did mathematics in his spare time but published little and proved even less. Many of his conjectures, especially in number theory, have since turned out to be true. Others are as yet unresolved.

*Conjecture 1*  $\forall$ natural numbers $n$, $2^{2^n} + 1$ is a prime number.
*Conjecture 2*  $\forall$ natural numbers $n$ which do not have $p$ as a prime factor, $n^{p-1}$ is divisible by $p$.
*Conjecture 3*  $\forall$ natural numbers greater than 2, $x^n + y^n = z^n$ has no integer solutions $x, y$ and $z$.

The first conjecture is false. Although $2^{2^1} + 1 = 5$, $2^{2^2} + 1 = 17$, $2^{2^3} + 1 = 257$ and $2^{2^4} + 1 = 65\,537$ are all prime (they are called *Fermat primes*), the conjecture fails for $n = 5$. This counterexample was produced by Euler in 1732:

$$2^{2^5} + 1 = 641 \times 6\,700\,417$$

It is not known whether there are any more Fermat primes.

The second conjecture, known as the 'Little Fermat Theorem', was made by Fermat in 1640. The proof was provided by Euler a century later.

The final conjecture is known as 'Fermat's Last Theorem'—whether it is true or not is unknown. The reader should notice that the equation $x^2 + y^2 = z^2$ does admit whole number solutions. The triples $(x, y, z)$ form the side lengths of a right-angled triangle (see Figure 1.7). The smallest solutions are the $(3, 4, 5)$ and $(5, 12, 13)$ triangles. The history of mathe-

**Fig. 1.7**

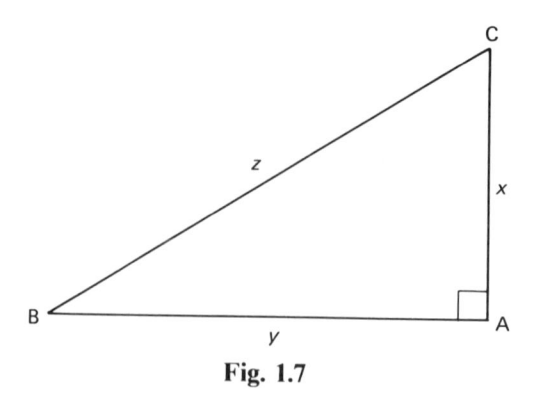

maticians' attempts to resolve Fermat's Last Theorem is an interesting one. Fermat himself showed that the result was true in the case $n=4$. That is, $x^4 + y^4 = z^4$ admits no integer solutions. Euler, in the next century, did the case $n=3$. By 1955, with the aid of the SWAC digital computer, the result was proven for $n < 4003$. The result has been verified for $n < 25\,000$. Offers of prize money for the proof for all natural numbers $n > 2$ has resulted in a deluge of alleged proofs by amateurs and professional mathematicians alike. Fermat's Last Theorem has the dubious distinction of being the mathematical theorem possessing the largest number of proofs—all incorrect! For more details of this fascinating problem see *The Last Problem*, by E. T. Bell.

We have omitted from this section a very powerful method of proof, namely mathematical induction. This has a whole chapter to itself (see Chapter 5). The method is ideally suited to proving theorems of the form '$\forall$ natural numbers $n$, $P(n)$'. In our discussion of predicates we have talked about $P(x)$, where $x$ lay in some particular collection of objects. We put this latter concept on a more formal footing in the next section.

## 1.3 SETS

A given collection of objects is called a *set* and the objects in the set are called its *elements*. We shall write $a \in S$ whenever $a$ is an element of a given set $S$. The symbol $\in$ can be regarded as abbreviating the phrase 'belongs to'. It is, in fact, the initial letter of the Italian word *este*, meaning 'is'. This was only discovered by the authors during the final stages of the production of this book! We also write $b \notin S$ whenever $b$ does not belong to the set $S$. If the elements of a set can be listed, we do so and contain them between braces. For example, the set $S$ whose two elements are the dynamic duo encountered by Alice in the forest is written as

$$S = \{\text{Tweedledee}, \text{Tweedledum}\}$$

Predicates can also be used to specify the contents of a set—they are essential when our sets are uncountable. In other words, the elements cannot be listed.

(See Chapter 9.) As an example of the use of predicates, consider the set $T$ in Figure 1.8. The jargon can be interpreted as 'the set $T$ consisting of all real numbers $x$ which are less than 2'.

$$T = \{x: x \text{ is a real number and } x < 1\}$$

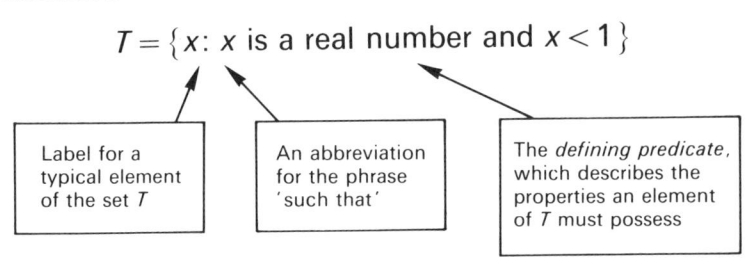

- Label for a typical element of the set $T$
- An abbreviation for the phrase 'such that'
- The *defining predicate*, which describes the properties an element of $T$ must possess

**Fig. 1.8**

When discussing any problem, conjecture or theorem in mathematics, we must always specify what set or sets are under consideration. This is most easily accomplished by specifying some *universal set*, $U$, for the problem. The sets which then arise only contain elements from this universal set. In this book we shall be examining, often in great detail, various sets of numbers. We denote these sets as in the following list—one of the objects of this book is to help you appreciate what exactly are the defining predicates in question!

The *natural numbers*:
$$\mathbb{N} = \{n: n \text{ is a positive whole number}\}.$$
The *integers*:
$$\mathbb{Z} = \{n: n \text{ is a whole number}\}.$$
The *rational numbers*:
$$\mathbb{Q} = \{p/q: p \text{ and } q \text{ are integers and } q \neq 0\}.$$
The *real numbers*:
$$\mathbb{R} = \{x: x \text{ is a real number}\}.$$
The *complex numbers*:
$$\mathbb{C} = \{a + ib: a \text{ and } b \text{ are real numbers and } i^2 = -1\}.$$

The definition of $\mathbb{R}$ in particular is clearly tautological—skip to Chapter 4 for a more satisfactory (?) explanation.

Manipulating the various sets which arise when tackling mathematical problems essentially disguises the fact that our arguments are deeply rooted in logic. Our logical connectives are buried in the definitions of the fundamental operations on sets, namely *union*, *intersection* and *complementation*. Similarly, the laws of the algebra of sets, which we shall shortly derive, embody logical equivalences of certain compound statements. We shall illustrate our set operations using Venn diagrams. These useful 'aides-memoire' are named after the British mathematician John Venn (1834–1923). First, we define the set-theoretic concepts needed.

A set $B$ is called a *subset* of a set $A$ if every element in $B$ is automatically an element of $A$. This is denoted by $B \subseteq A$ and is logically interpreted as 'If $x \in B$, then $x \in A$'. The corresponding Venn diagram is Figure 1.9.

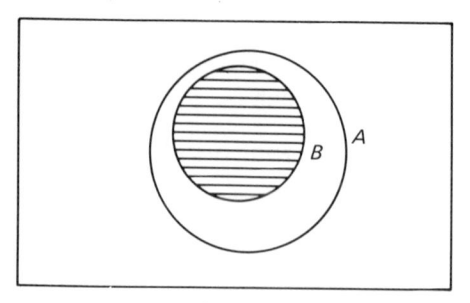

**Fig. 1.9**

Two sets are *equal*, written as $A = B$, provided that they have precisely the same elements. To prove that two given sets are equal requires us to show that each is a subset of the other. In other words, $A = B$ if and only if $A \subseteq B$ and $B \subseteq A$. For an instructive example, see Section 4.4. One rather trivial consequence of set equality is that the elements of a given set need not appear in any particular order.

The *union* of two sets $A$ and $B$ is the set of all objects belonging to either (or both) of the two sets. The notation is $A \cup B$ and formally $A \cup B = \{x : x \in A \vee x \in B\}$. (See Figure 1.10).

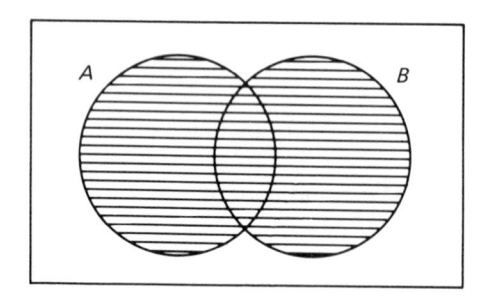

**Fig. 1.10**

The *intersection* of two sets $A$ and $B$, denoted by $A \cap B$, is the set of all elements common to both $A$ and $B$. In set-theoretic language $A \cap B = \{x : x \in A \wedge x \in B\}$. (See Figure 1.11).

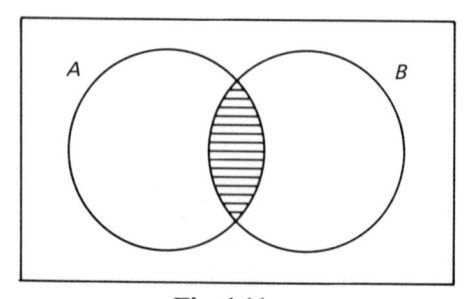

**Fig. 1.11**

The *complement* of a set $A$ is the set of all elements (in the universal set) not in $A$. The notation that we shall use is $A'$. So $A' = \{x : x \notin A\}$. (See Figure 1.12).

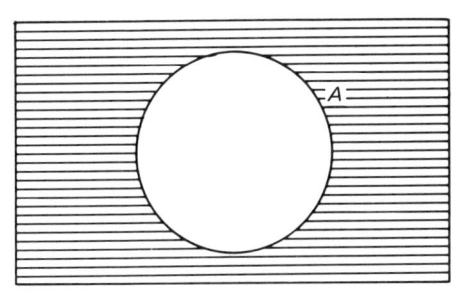

**Fig. 1.12**

In our Venn diagrams the rectangle represents the universal set $U$ within which we are working. Elements of the sets $A$ and $B$ are within the appropriately labelled circles, and the shaded region indicates the set combination being depicted.

We now deduce from our basic definitions certain fundamental properties of the algebra of sets. These appear in Figure 1.13, where $\emptyset$ denotes the

| Associative laws: | |
| --- | --- |
| $A \cup (B \cup C) = (A \cup B) \cup C$ | $A \cap (B \cap C) = (A \cap B) \cap C$ |
| **Commutative laws:** | |
| $A \cup B = B \cup A$ | $A \cap B = B \cap A$ |
| **Identity laws:** | |
| $A \cup \emptyset = A$ | $A \cap U = A$ |
| $A \cup U = U$ | $A \cap \emptyset = \emptyset$ |
| **Idempotent laws:** | |
| $A \cup A = A$ | $A \cap A = A$ |
| **Distributive laws:** | |
| $A \cap (B \cup C) = (A \cap B) \cup (A \cap C)$ | $A \cup (B \cap C) = (A \cup B) \cap (A \cup C)$ |
| **Complement laws:** | |
| $A \cup A' = U$ | $A \cap A' = \emptyset$ |
| $U' = \emptyset$ | $\emptyset' = U$ |
| $(A')' = A$ | $(A')' = A$ |
| **De Morgan's laws:** | |
| $(A \cup B)' = A' \cap B'$ | $(A \cap B)' = A' \cup B'$ |

**Fig. 1.13**

*empty set*, the set with no elements. This set arises naturally if, for example, we are taking the intersection of two sets which have no elements in common. Notice that $\emptyset \subseteq A$ for any set $A$, since every element of $\emptyset$ is, by default, an element of $A$!

The reader is invited to prove these laws for him/herself. As an illustration we establish that $(A \cup B)' = A' \cap B'$:

$$\begin{aligned} x \in (A \cup B)' &\Leftrightarrow x \notin A \cup B \\ &\Leftrightarrow \sim (x \in A \cup B) \\ &\Leftrightarrow \sim (x \in A \vee x \in B) \end{aligned}$$

Now $\sim (P \vee Q)$ is logically equivalent to $\sim P \wedge \sim Q$ when $P$ and $Q$ are statements. An obvious extension to predicates $P(x)$ and $Q(x)$ enables us to use truth tables to establish that $\sim (P(x) \vee Q(x))$ is logically equivalent to $\sim P(x) \wedge \sim Q(x)$. We now let $P(x)$ denote '$x \in A$' and $Q(x)$ denote '$x \in B$'. Hence,

$$\begin{aligned} x \in (A \cup B)' &\Leftrightarrow \sim (x \in A \vee x \in B) \\ &\Leftrightarrow x \in A' \wedge x \in B' \\ &\Leftrightarrow x \in A' \wedge B' \end{aligned}$$

In Figure 1.13 the laws in any one column are called the *duals* of the ones in the other column. Notice that any law can be turned into its dual by interchanging $U$ with $\emptyset$ and swapping $\cap$ with $\cup$. The complement symbol $'$ is left alone, which explains the repeat in our table of the self-dual law $(A')' = A$. Now any identity involving sets which can be deduced from our laws gives rise to a dual identity which can be obtained by interchanging symbols as detailed above. For example, the identity $A \cap (A \cup B)' = \emptyset$, [**1.6**], has dual $A \cup (A \cap B)' = U$. This *principle of duality* occurs frequently in mathematics, notably in the study of projective geometry.

Once all the laws of the algebra of sets have been proved, they can be used to derive more sophisticated relationships between sets. The proofs are slicker and our arguments do not appear so pedantic as, for example, when we proved one of De Morgan's laws. This is natural, since in one sense we can assume more. Where it all begins (in other words, what are our fundamental assumptions?) is a matter of debate, and to a certain extent we have some choice. One could view the laws of the algebra of sets as our fundamental assumptions. The reader is invited to speculate whether there are certain set-theoretic results which require us to descend to a more basic level of argument than our laws (apart from the laws themselves). Setting up axiom systems is a dangerous business. Have we left any necessary assumptions out? Do any of our axioms contradict others? One approach to an understanding of the set of real numbers, $\mathbb{R}$, is to specify axioms which our set of real numbers shall satisfy. From these we hope to deduce all the familiar properties of the real numbers. More importantly, perhaps we can gain a better understanding of what real numbers actually are.

As an early introduction to the deduction of theorems from a set of axioms, we use the algebra of sets to prove that the set

$X = (A \cap B \cap C) \cup (A' \cap C) \cup (B' \cap C)$ is just the set $C$ in disguise:

$$\begin{aligned}
X &= (A \cap B \cap C) \cup ((A' \cap C) \cup (B' \cap C)) && \text{using an associative law} \\
&= (A \cap B \cap C) \cup ((C \cap A') \cup (C \cap B')) && \text{using a commutative law} \\
&= (A \cap B \cap C) \cup (C \cap (A' \cup B')) && \text{by a distributive law} \\
&= (A \cap B \cap C) \cup (C \cap (A \cap B)') && \text{via De Morgan's law} \\
&= (C \cap (A \cap B)) \cup (C \cap (A \cap B)') && \text{by associativity and commutativity} \\
&= C \cap ((A \cap B) \cup (A \cap B)') && \text{using a distributive law} \\
&= C \cap U && \text{by a complement law} \\
&= C && \text{by an identity law}
\end{aligned}$$

Another operation on sets that we shall need in Chapters 3 and 9 is the *Cartesian product*, $A \times B$, of two sets $A$ and $B$:

$$A \times B = \{(a,b): a \in A \wedge b \in B\}$$

Elements of this set can be regarded as ordered pairs where the first component lies in the set $A$ and the second in the set $B$. In particular, notice that this operation is not commutative. In fact, when is $A \times B = B \times A$? This operation does not mix well with the previous ones. Show that $(A \times B) \cap (B \times A) = (A \cap B) \times (A \cap B)$, but that $(A \times B) \cup (B \times A)$ is, in general, only a proper subset of $(A \cup B) \times (A \cup B)$. [**1.7**]

Other operations on sets can be defined in terms of previous defined operations. A couple of examples, of some independent interest, are the *difference*, $A \setminus B$, and *symmetric difference*, $A \triangle B$, of sets $A$ and $B$:

$$A \setminus B = \{x : x \in A \wedge x \notin B\}$$
$$A \triangle B = \{x : (x \in A \cup B) \wedge (x \notin A \cap B)\}$$

Draw the Venn diagrams of these two sets. [**1.8**]

# 2 Unique Factorisation—in which trivial arithmetic reveals a glimpse of hidden depths

'Arithmetic is one of the oldest branches of human knowledge; and yet some of its most abstruse secrets lie close to its tritest truths.'

H. J. S. Smith

'There still remain three studies suitable for free man. Arithmetic is one of them.'

Plato

This chapter concerns mainly familiar old friends, the set $\mathbb{N}$ of positive whole numbers variously known as the natural numbers or counting numbers, $\{1, 2, 3, 4, \ldots\}$. We think of counting as a very primitive notion firmly rooted in reality, yet already the innocent three dots in $\{1, 2, 3, 4, \ldots\}$ may have taken us beyond reality into the realms of pure thought. The dots are usually interpreted as 'and so on for ever', which expresses our notion that $\mathbb{N}$ is an *infinite* set. Cosmologists have still not made up their minds whether we live in a finite or an infinite universe, and in the former case there could be no such thing as an infinite set of real objects. However, mathematicians do not, in general, see this as a problem, and confidently assume that infinite sets (even if they are only sets of *mental* objects) can be handled with safety.

There will be much to say about infinity later, but for the moment we concentrate on another fairly primitive notion—that of splitting a natural number into its prime factors. Consider, for example, the number 10 780, and suppose that in a fit of enthusiasm I decide to find all the prime numbers not exceeding half of 10 780 and then find which combinations of them have a product of 10 780. If just a little of this enthusiasm is replaced by thought, I may carry out a few divisions to arrive at

$$
\begin{array}{r|r}
2 & 10\,780 \\
2 & 5390 \\
5 & 2695 \\
11 & 539 \\
7 & 49 \\
7 & 7 \\
& 1
\end{array}
$$

# UNIQUE FACTORISATION

from which I conclude that $10\,780 = 2^2 \times 5 \times 7^2 \times 11$, and the job is done.

I feel some regret that two days' solid work went into the first method before I realised that two minutes' work would suffice, but at least the two methods agreed, so I also have a check on my work. I can derive further reassurance that my two days' hard labour was not pointless from the observation that the long method will in principle find all families of primes whose product is a given number, whereas the quick method only produces one. Admittedly, there is only one family of primes whose product is 10 780, but for very big numbers, for which the number of trial factors is enormous, it is by no means obvious that this will always be the case. You may, of course, feel intuitively that each whole number has a unique prime factorisation, and in this case your intuition would be reliable. But so many equally strong intuitions held by quite brilliant mathematicians have proved to have been misguided in the past that complacency based on strength of intuition must be suspect. [We need to distinguish between *sets*, which by definition have all their elements distinct, and *families*, which can have repeated elements. Thus, the families $\{2, 2, 5, 7, 7, 11\}$ and $\{2, 2, 2, 5, 7, 11\}$ are different but their *sets* of elements are the same, namely $\{2, 5, 7, 11\}$.]

Our theorem, which we shall refer to as Uniqueness of Prime Factorisation (UPF), sometimes called the Fundamental Theorem of Arithmetic, is precisely that, given any natural number $n$, there is only one family of primes whose product is $n$. [We also use UPF for 'unique prime factorisation' and for the rather clumsy phrase, 'uniquely prime factorisable'. Which of these is intended will be clear from the context.]

C. F. Gauss was the first to find a proof (in 1801), and, although it was quite subtle, his main contribution was to realise that a proof was necessary at all. Several mathematicians had done quite profound work in the theory of numbers before this date and had probably just not noticed that many of their results and procedures depend upon UPF for their validity. Carl Friedrich Gauss (1777–1855) may well have spent his life as an obscure bricklayer but for a perceptive teacher who recognised his ability at an early age. He was a genius in two distinct ways: first, he contributed significantly to an enormous range of subjects stretching from astronomy and electricity to probability, number theory and geometry; perhaps more importantly, he showed amazing insight into the nature of mathematics, which enabled him to be a link between the intuitive eighteenth century and the rigorous nineteenth. His perception of UPF is just one example of this. Another is his anticipation that the axioms of geometry were rather more than translations into mathematical terms of known truths about physical space, and that other consistent geometries could exist.

Rather than just plunge into the proof, let us take a rather more roundabout route through scenery against which UPF will stand out as a very odd result. The point of the journey is motivational—to make you *want* to struggle with the details of the proof. After all, it must be admitted that mathematics is hard work, and although solving hard problems is more satisfying than solving easy ones, we do need the conviction that the effort is worth while.

First, then, a fairly mundane point—that we agree to define a prime number as one with exactly two factors. This neatly excludes 1, admittedly only by convention but nevertheless a very useful convention, without which UPF would plainly be false ($2 \times 3, 1 \times 2 \times 3, 1^2 \times 2 \times 3, \ldots$ would be *different* prime factorisations of 6). We are thereby classifying the members of $\mathbb{N}$ into three mutually exclusive subsets, 1—in a class of its own; the primes; and the rest, called composite numbers. There are many conventional definitions in mathematics and it is important to realise that they are only introduced to make it easier to talk about the results we prove. They don't change the results! Humpty Dumpty certainly had a point when he told Alice that he could use a word to mean whatever he chose, but only if he was indifferent to his ability to communicate with others.

Next we move to an important consequence of UPF. Suppose that you require the highest common factor of two numbers, say 10 780 and 39 102. One method of doing this is to factorise both, then pick out the highest power of each prime common to both factorisations, and multiply them all together: $10\,780 = 2^2 \times 5 \times 7^2 \times 11$, $39\,102 = 2 \times 3 \times 7^3 \times 19$, so the highest common factor is $2 \times 7^2 = 98$. Clearly, 98 will be *a* common factor of both but only with UPF can we guarantee that it is the biggest. Why? [**2.1**]

The most interesting part of our approach to UPF comprises a close scrutiny of a number system $S$ which, as far as the ideas of prime and factorisation are concerned, seems indistinguishable from $\mathbb{N}$, yet UPF does not hold in $S$. $S$ is just a subset of $\mathbb{N}$ defined as $\{4n - 3 : n \in \mathbb{N}\}$, so $S = \{1, 5, 9, 13, 17, 21, 25, \ldots\}$. $S$ has the property of being closed under multiplication—if $s$ and $t$ are in $S$, then so is $st$. Prove this. [**2.2**] This means that we can give a three-way classification of $S$ just as we split $\mathbb{N}$ into $\{1\}$, the primes and the composites. The three classes in this case are $\{1\}$, those $S$ members $s$ which have exactly two factors in $S$ (those $S$-factors being just $S$ and 1), and the rest. It makes sense to call those in the second class the *S-primes*. Clearly, anything in $S$ which is prime is necessarily an $S$-prime (for example, 17), but there are members of $S$ such as 21 which are $S$-prime but not prime. Now consider the number 1617. It is in $S$ [**2.3**], and $1617 = 33 \times 49 = 77 \times 21$, with $33, 49, 77$ and 21 all being $S$-primes [**2.4**], so we have an example to show that the system $S$ does not satisfy a unique $S$-prime factorisation theorem.

As you may have guessed, 1617 wasn't just found by luck. If you think about it, you should be able to find lots of $S$-numbers which are not uniquely factorised into $S$-primes. [**2.5**] But is this any cause for surprise? After all, $S$ and $\mathbb{N}$ are not the same, so why should we expect them to have the same properties? We could invent several systems like $S$ and it would be reasonable to say that some of them will obey UPF and others won't, and that's all there is to it. Such a view *is* reasonable, so why are we not prepared to leave it at that? One answer is that facts may be interesting but they are only the start of the story. A characteristic of mathematicians is their feeling that their knowledge is seriously incomplete if they only know *that* something is the case without understanding *why* it is so. In the context of our present example, this means that they have an urge to dig deeper and ask exactly which differences between ℕ and S are significant from the point of view of UPF. It is as a result of this digging that the UPF property of ℕ shows up as *surprising*.

Let us see which of the many features of ℕ we actually need in order to define and talk about UPF. We are concerned with factors and primes, which clearly have something to do with multiplication, and indeed multiplication is *all* we need. It is worth spelling this out in detail: in ℕ, $a$ is called a factor of $b$ if $b = ka$ for some natural number $k$, with a natural number being prime if and only if it has exactly two distinct factors, and ℕ is closed under multiplication. Now go back over the previous sentence, replacing ℕ by S, factor by S-factor, natural number by S-number, prime by S-prime, and you will have analogous definitions and true statements in S.

Since we know that UPF does not hold in S, we also know that if we succeed in proving UPF for ℕ, something must go wrong if we just 'change the words' in this proof from ℕ-words to S-words. Our previous remarks strongly suggest that what goes wrong will have nothing to do with multiplicative properties. We give a proof below, but before reading this, it is worth having a guess about the significant difference between ℕ and S. [**2.6**]

## A PROOF OF UPF FOR ℕ

### Outline of Proof

*Step 1* Suppose that there is a natural number having more than one PF (in other words, UPF is *not* true), and let $k$ be the smallest such number.

*Step 2* Deduce from this that a *smaller* natural number exists with the same property. This clearly contradicts the definition of $k$, so there is no smallest non-UPF natural number, and, hence, no non-UPF natural number at all.

In order to get from step 1 to step 2 we need two bridges, B1 and B2. These are:

B1: If $k$ (the number introduced in step 1) has PFs $pqr...$ and $p'q'r'...$, then the families $(p, q, r, ...)$ and $(p', q', r', ...)$ are disjoint.

B2: If $n$ is UPF and its UPF is $p_1 p_2 ... p_l$, then $n$ has no prime factor outside the family $(p_1, p_2, ..., p_l)$.

The scheme for our proof appears in Figure 2.1.

Before filling in the details of this scheme, we pause to consider feelings you may have about B2. Isn't it obvious?—and why is it significant? Suppose, for example, $n = 2 \times 5 \times 7^3 \times 53 \times 67$, and we know that $n$ is UPF. Is it possible that a prime other than 2, 5, 7, 53 or 67 could be a factor of $n$—say the prime 13? No, because if it was, we could then factorise $n$ as 13 × some

```
        B1          B2
Step 1  ────→  •  ←────  Step 2
        ←──────────────→
            demolishes
```
**Fig. 2.1**

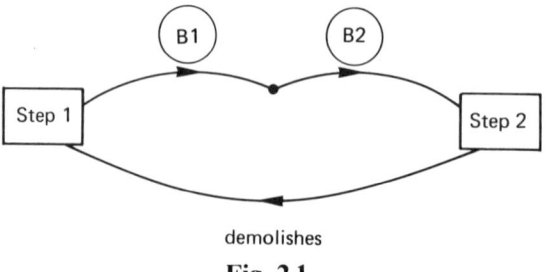

product of primes, so the PF of *n* would not be unique (we started with one not containing 13 and produced another which does!) In this sense B2 is obvious. As for its significance, you will see clearly how it is used in the proof which follows shortly.

To see that B1 is true, if $p$ was in both families (for example, if $p_1 = p_1'$), then $qr\ldots$ and $q'r'\ldots$ would be different PFs of $k/p$, a number *smaller* than $k$, which is impossible, because $k$ is the *smallest* natural number with non-UPF.

## The Main Proof

We are free to write the primes in the factorisation of $k$ in any order, so for convenience we write them in non-decreasing order. That is, $p \leqslant q \leqslant r \leqslant \ldots$ and $p' \leqslant q' \leqslant r' \leqslant \ldots$.

Both lists contain at least two members. Why? [**2.7**] From this we see that $k \geqslant p^2$ and $k \geqslant p'^2$, and since by B1 $p \neq p'$, we have $k > pp'$.

Hence, $k - pp'$ is a natural number less than $k$ and divisible by both $p$ and $p'$. So, by the minimal defining property of $k$ and by B2, the UPF of $k - pp'$ must be of the form $pp'QR\ldots$ (or perhaps just $pp'$), where $Q, R, \ldots$ are further primes.

So
$$k = pp' + pp'QR\ldots$$

That is,
$$pqr\ldots = pp' + pp'QR\ldots$$

So
$$qr\ldots = p' + p'QR\ldots$$

and we see that $p'$ is a factor of $qr\ldots$.

But $p' \notin (q, r, \ldots)$, because by B1 the families $(p, q, r, \ldots)$ and $(p', q', r', \ldots)$ are disjoint. So, by B2 (or strictly, its contrapositive), the number $qr\ldots$ is non-UPF. This is a contradiction we require to finish the proof, because $qr\ldots < k$.  □

Now can you identify a property of $\mathbb{N}$ which was required in this proof but which does not hold for $S$? [**2.8**]

Although the proof above was presented as one of the contradiction type, it is really induction in disguise. If you have previously met induction only in the form 'Verify result for $n=1$, show that the result for $n+1$ follows from the result for $n$', the disguise may seem rather good! If your immediate interest is in seeing through the disguise, turn to Chapter 5. Chapter 6 contains a totally different sort of unique factorisation result having nothing to do with numbers. Highest common factors can be found without making any appeal to UPF and this is pursued in Section 4.5.

Finally, to return to the theme of this chapter, here is another number system, rather different from $S$, in which UPF fails. Define $T$ as the set of all complex numbers of the form $a + b\sqrt{-6}$, where $a$ and $b$ are integers. Then split $T$ into three disjoint subsets: the 'specials' $\{0, 1, -1\}$; the $T$-primes, which are those members of $T$ other than the specials which cannot be written as the product of two non-special members of $T$; and the rest, which we may as well call $T$-composites. To gain some familiarity with $T$, you could start by finding a prime which is not a $T$-prime [**2.9**], then show that there is more than one way of factorising 10 into $T$-primes. [**2.10**] Can you spot a property of $\mathbb{N}$ used in the proof of UPF which does not hold in $T$? [**2.11**] (UPF fails in $S$ and $T$ for *different* reasons.)

## SOME ROUTE CHANGES

In Chapter 3 our first aim (Section 3.1) is to indicate how our progression in understanding from the natural numbers to the positive and negative fractions depends not just on our ability to cope with generalisation, but also on carrying as part of our mental luggage several interpretations of numbers which are, in a sense, *contradictory*. It is likely to be of most interest to those intending to teach mathematics, and is independent of Sections 3.2 and 3.3, which take up the purely logical development.

Section 3.4 is about the very important idea of an equivalence relation, which is used in Chapters 4, 6 (marginally) and 9.

Chapter 4 contains what we consider the most important theme of the whole book—the real numbers. We begin the journey from the rationals to the reals by taking for granted only an intuitive understanding of fractions, so that this chapter can be read independently of all of Chapter 3 except Section 3.4. The exercises from [3.31] onwards are not essential.

So the choices now available to you are:

(1) Continue with 3.1.
(2) Omit 3.1 but read the rest of Chapter 3 } with or without
(3) Omit 3.1, 3.2 and 3.3 and read 3.4      } exercises [**3.31**] to [**3.37**].
(4) Omit all of Chapters 3 and 4 and continue with Chapter 5 or 6—but Chapters 7 onwards require Chapter 4.

# 3 Numbers—in which we abandon logic to achieve understanding, then use logic to deepen understanding

'God made the natural numbers, all else is the work of man.'
Leopold Kronecker (1823–1891)

'"Can you do Addition?" the White Queen asked. "What's one and one and one and one and one and one and one and one and one and one?"
"I don't know," said Alice. "I lost count."
"She can't do Addition," the Red Queen interrupted.'
Lewis Carroll

## 3.1 'Sums' in School

How and why do we abandon logic while we learn about numbers? Just think back to your first meeting with 'one', 'two', 'three'.... These characters were probably of very minor importance initially, being no more than labels to distinguish the verses of 'one, two, buckle my shoe', etc. Then they began to have relationships with one another—'one' for some reason always came before 'two'; and relationships with the outside world—'one', 'two', 'three' were names attached to small collections of things, but 'ten', 'eleven', 'twelve' were associated with bigger collections. So, even at pre-school age, we had experience of numbers being used in at least three ways—as arbitrary labels, as a means of ordering events (Monday first, Tuesday second...) and as measures (of height, weight, more numerous, less numerous...)—and these interpretations seem to have very little to do with one another. If we imagine pre-school children sufficiently precocious to ask 'What *is* a number?' it would be extremely difficult to give them a sensible answer.

At school we begin to combine numbers together in various ways—by adding, subtracting, multiplying and dividing them—and these operations have to be made meaningful and their results reasonable. The difficulties for teacher and pupil now begin to mount up. If we consider just one of these operations (say, addition), it is clear that if numbers are interpreted in the 'how many?' sense, then Figure 3.1 will serve as an interpretation of $3 + 2 = 5$,

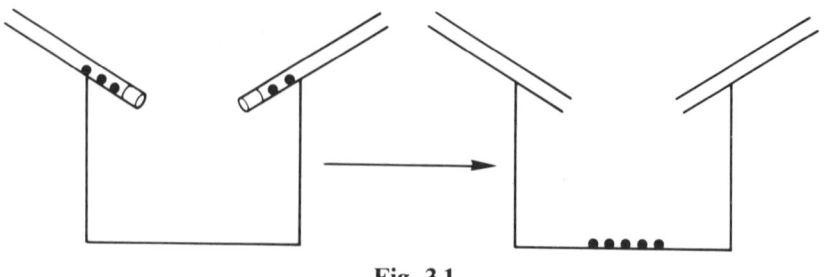

**Fig. 3.1**

where the + sign is an instruction to lump two collections together or form the union of two disjoint sets.

But what if we are thinking of numbers in their ordinal sense? If you were second in the race and I was third, there isn't much sense in the claim that together we were fifth! These rather trivial examples only serve to show that if we want $3 + 2 = 5$ to mean anything in the real world, we have a choice in the interpretation of numbers. In some interpretations $3 + 2 = 5$ makes sense; in others it doesn't. It is also worth pointing out that tipping tubes of beads into a beaker makes fundamental properties of addition such as $3 + 2 = 2 + 3$ and $(3 + 2) + 6 = 3 + (2 + 6)$ obvious.

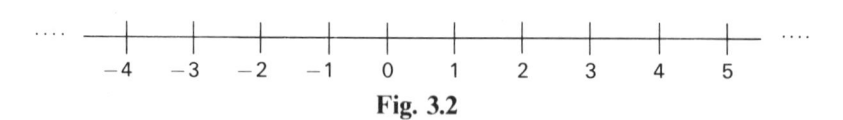

**Fig. 3.2**

Another commonly used model which is especially useful for introducing the integers is the number line, represented as in Figure 3.2. The numerals seem to be standing for the positions of a set of evenly spaced points on this line, but if this is really what is intended, how are we to make sense of addition? It was reasonable to think of $3 + 2 = 5$ as symbolising the amalgamation of two sets of beads to produce a larger set, but how are two *positions* on the number line to be combined to produce another position? It just can't be done in any natural way; and if we want to use the idea of a number line to model addition, we have to change interpretation yet again. One strategy is to say that $3 + 2$ means 'Start at 3 and take two steps to the right'. The answer, 5, is then the label of the end-point of this process. This works, but the interpretation is rather clumsy, in that the 3 is a 'where?' number (the position of a point) and the 2 is a 'how many?' number, and we lose the naturalness of $3 + 2 = 2 + 3$. A model not open to this objection is to interpret all integers in the sense of 'how many steps?' and then

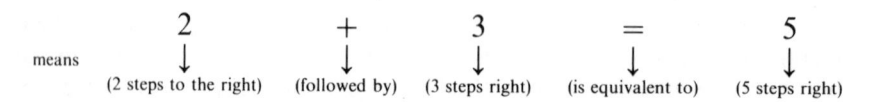

| means | 2<br>↓<br>(2 steps to the right) | +<br>↓<br>(followed by) | 3<br>↓<br>(3 steps right) | =<br>↓<br>(is equivalent to) | 5<br>↓<br>(5 steps right) |

and $2 + {}^-3 = {}^-1$ asserts the equivalence of two steps right followed by three steps left with one step left, so we have a decent model for adding any two integers. Notice that the model is independent of the origin: $2 + 3 = 5$ says that we shall end up five steps to the right of our starting point, *wherever that was*. So, if we want to retain the imagery of Figure 3.2, we can think of the numerical labels on the diagram as labelling the *number of steps* required to reach that point from 0.

When we come to subtraction, there are several choices again. The $-$ in $2 - 3 = {}^-1$ can be thought of as 'followed by the reversal of', so that $2 - 3 = 2 + {}^-3$ and subtraction has been defined (or explained away?) in terms of integer addition. Or we can use the idea of subtraction being the inverse of addition, so that asking $2 - 3 = ?$ is equivalent to asking $3 + ? = 2$.

For multiplication 'three lots of two' will do for $3 \times 2$, but ${}^-3 \times 2$, $3 \times {}^-2$ and ${}^-3 \times {}^-2$ require some pretty radical rethinking. One neat way of modelling multiplication is to think of the numbers involved as speed factors, as in Figure 3.3(a), (b), where the left-hand wheel is driving the right-hand

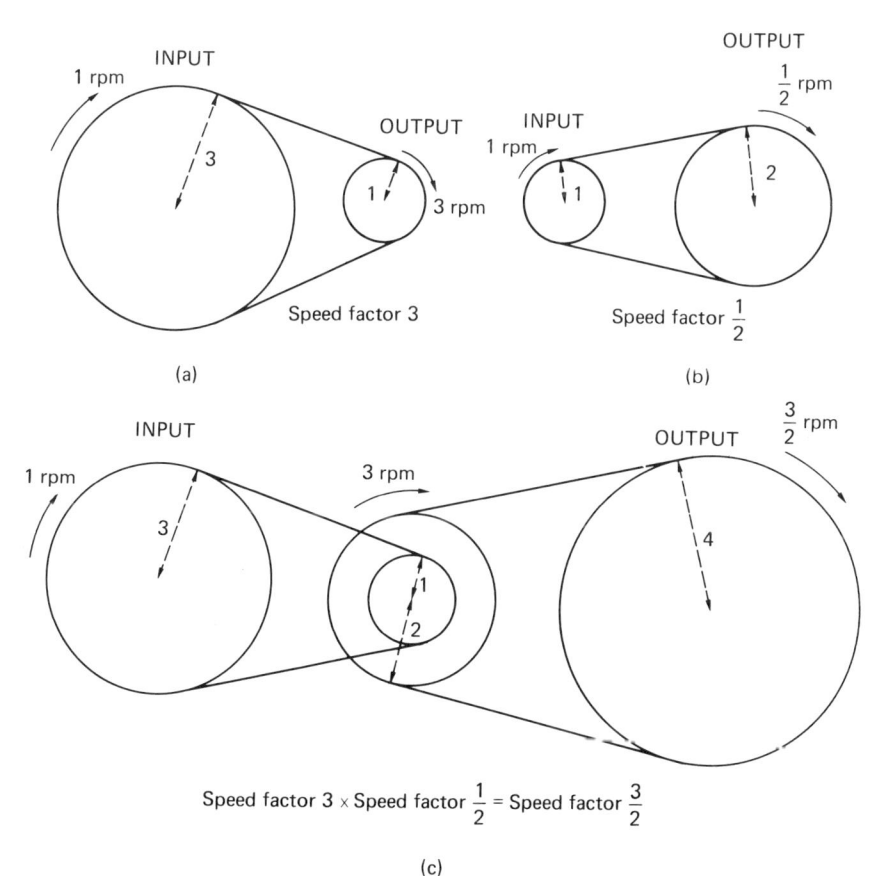

**Fig. 3.3**

INPUT                                                    OUTPUT

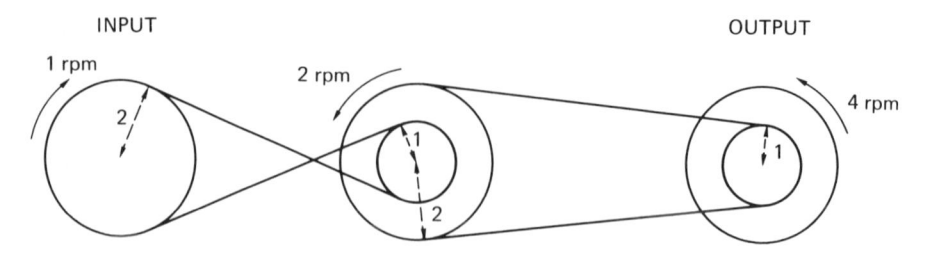

1 rpm                    2 rpm
                                                                    4 rpm

Speed factor 2 in reverse × Speed factor 2 = Speed factor 4 in reverse

**Fig. 3.4**

one by a non-slipping belt. When two such systems are coupled, as in Figure 3.3(c), the net speed ratio of the combined machine is obviously the product of the two separate component speed ratios. Negative numbers are dealt with by simply crossing the driving belt, as in Figure 3.4.

For division $15/3$ and $^-15/3$ can be pictured as sharing a profit or loss of £15 between three shareholders, but that sort of model can't make sense of $15/^-3$, or of $15/(14/11)$.

Our final model of addition and multiplication of numbers returns to the number line. If $x$ and $y$ are given numbers on the number line, the method of fixing the position of $x+y$ is to imagine arrows for $x$ and $y$, as in Figure 3.5(a). $x+y$ is defined as the position of the 'sharp end' of $y$'s arrow after sliding it along the line until its 'blunt end' coincides with the sharp end of $x$, as in Figure 3.5(b). This works for any combination of signs of $x$ and $y$.

The *definition* of $^-x$ can also be incorporated as part of the specification of the model: if we have only defined the position of zero and the positive

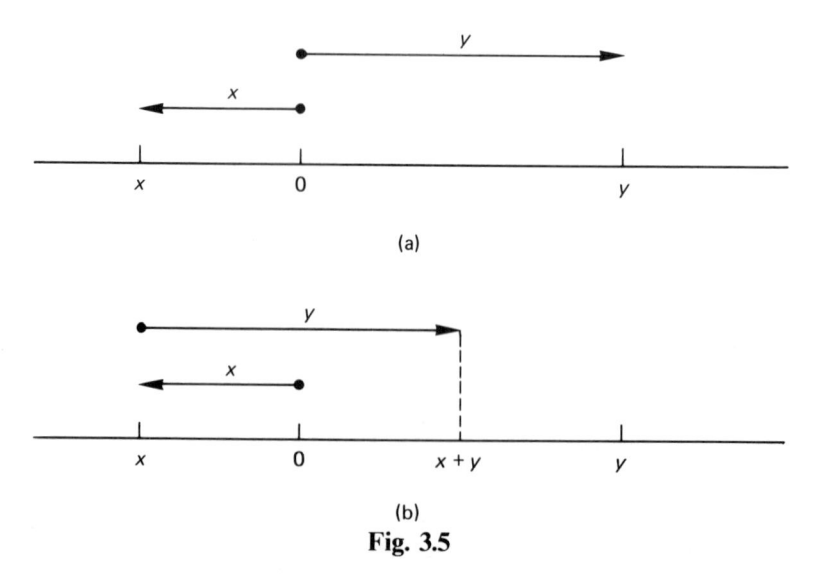

(a)

(b)

**Fig. 3.5**

$x$'s, $^-x$ can be defined as the sharp end of $x$'s arrow after giving it a half-turn around its blunt end (zero). In this model such properties as $x + 0 = x$, $x + y = y + x$, $^-(x + y) = (^-x) + (^-y)$ and $^-(^-x) = x$ all become fairly obvious.

The same model will cope with subtraction either by interpreting $x - y$ as $x + (^-y)$ or by defining $x - y$ as the solution for $c$ of the addition equation $y + c = x$. You should check that you can operate the model to yield $x - y$ using either of these interpretations. [**3.1**]

For multiplication we require two number lines, a fixed rigid reference line, $L$, and another which consists of an ideal spring, $S$. $S$ is firmly fastened down at one point, which is marked zero, and lined up so that in its unstretched and uncompressed state its numbers coincide with those on $L$. $S$ can be given a half-turn about 0, but the pivot itself is not allowed to move. In the case illustrated, the position of $xy$ on $L$ is found by the following procedure. Stretch $S$ until its 1 is lined up with $x$ on $L$. $xy$ is then the new position of $y$, as in Figure 3.6.

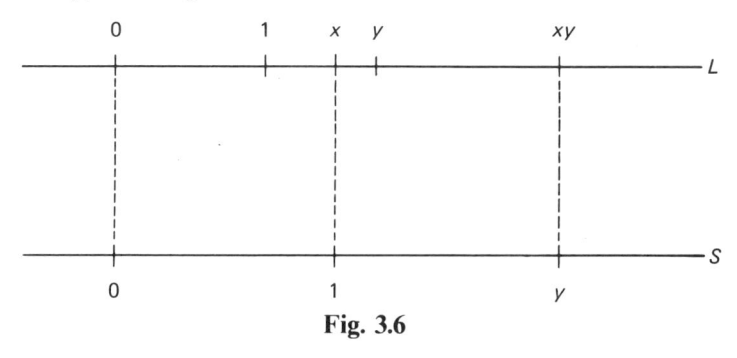

**Fig. 3.6**

Had $x$ been negative we would first reverse the direction of $S$ by giving it a half-turn about 0, then stretch or compress it until its 1 was in line with $x$ on $L$. The point on $L$ which was then in line with the $y$ on $S$ would be the product $xy$.

Such results as $x \times 1 = x$, $x \times 0 = 0$, $(^-x) \times (^-y) = xy$ and $(^-x) \times y = ^-(xy)$ should now be obvious on the basis of this model. [**3.2**]

To find $1/x$ we proceed as follows: stretch (or compress, or combine a stretch/compression with a half-turn) $S$ until its $x$ is lined up with the 1 on $L$. The position of 1 on $S$ then corresponds to that of $1/x$ on $L$. Why? [**3.3**] You should also be able to use this model to interpret the fact that there is no such number as $1/0$. [**3.4**]

To cope with the basic rule which connects the two operations of addition and multiplication, namely $z(x + y) = zx + zy$, we appeal to the *linear* nature of our ideal spring—that is, when we stretch or compress it, all gaps between points change by the same factor. So, if we do the stretch corresponding to multiplication by $z$, we have the situation shown in Figure 3.7.

Hence $zy = z(x + y) - zx$, and by our model of subtraction this means that $zx + zy = z(x + y)$.

**Fig. 3.7**

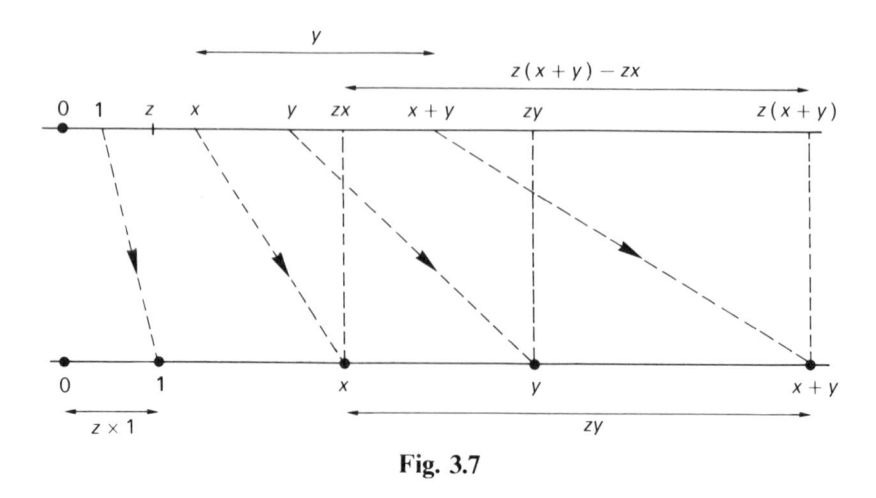

From a relatively sophisticated vantage point this model is neat, but the view from any lower level cannot be fully satisfying, simply because understanding linearity and making it plausible for real objects such as springs is probably harder than understanding multiplication!

We have now scratched sufficiently at the surface of this subject to realise how difficult it is to give a coherent satisfying development of numbers and of operations upon them. Different types of numbers and different operations seem to require different mental pictures to explain them. Furthermore, the pictures are often incompatible. What, for example, are we to make of $(^-6/5) \times (^-2 + 5/3)$ if addition corresponds to walks along the number line, division to sharing and multiplication to combinations of wheels and driving belts? This is where the illogicality in the title of this chapter comes in: in order to teach children about our number system, keeping in step with their normal experience of the world and with their psychological development, it seems necessary to make discontinuous conceptual jumps. It is a reasonable bet that good teachers will hope that their bright fifteen-year-old pupils will confidently manipulate $(^-6/5) \times (^-2 + 5/3)$, but will also hope that they will not ask too many searching questions about what it means!

## 3.2 BACK TO SQUARE ONE

It is only with hindsight that we are able to give a fully coherent account of all the numbers we have met — natural numbers, integers and fractions — and it is the power of abstraction which gives us the means of doing this. For example, $(^-6/5) \times (^-2 + 5/3)$ will turn out to be the same as $2/5$, not because of any particular interpretation in which the calculation arises but as a logical consequence of very few, very simple, very general laws which all our numbers obey.

One way in which this can be done is to take the natural numbers for granted (we have to start somewhere), and then to construct the integers, ensuring that integers and operations on them are defined entirely in terms of natural number operations. Then we construct the rationals from the integers, ensuring that their operations are defined entirely in terms of integers.

Our plan is to sketch the construction of the set of integers ($\mathbb{Z}$) from the natural numbers in this section, leaving you with 3.3 as a 'do it yourself' section to fill in some of the gaps and to continue with the next step, that of constructing the set of rational numbers ($\mathbb{Q}$) from the integers.

In a sense this chapter has the nature of a tidying-up operation. We are taking our previous hard-won understanding of numbers and showing how it can all be built up from the natural numbers, unaided by the mental props of interpretative models. Since we are polishing the logical structure of a body of facts which is already 'known', rather than creating new knowledge, we must not expect exciting revelations. A warm glow of satisfaction is a more realistic expectation. More striking surprises come in Chapter 4 in making the step from $\mathbb{Q}$ to the real numbers ($\mathbb{R}$), the step which is forced upon us as a consequence of the intuitions we have about continuity but which takes us way beyond anything with which intuition could cope without the assistance of pure mathematical inventiveness. This is not to say that the present chapter is devoid of importance, since it includes concepts of significance elsewhere in this book and more generally throughout mathematics—those of an equivalence relation and of a well-defined operation.

### 3.2.1 The Construction of $\mathbb{Z}$ from $\mathbb{N}$

To make a start, consider the negative whole numbers. We find a need for these, because we want to be able to talk about losses as well as gains, about temperatures below as well as above freezing, and about positions on the number line to the left as well as to the right of the origin. What is more significant for our purely logical purpose is that $\mathbb{N}$ is closed under addition but not subtraction, whereas $\mathbb{Z}$ is closed under both. In fact, any integer can be represented as the difference between two natural numbers ($^-5 = 7 - 12$, $0 = 2 - 2$, $3 = 8 - 5...$), so we represent these three integers by the *ordered pairs of naturals* (7, 12), (2, 2) and (8, 5), respectively. We shall attempt to use elements of $\mathbb{N} \times \mathbb{N}$ for $\mathbb{Z}$, with suitable operations to correspond to addition and multiplication. Now what guides us unofficially in seeking official definitions of addition and multiplication in $\mathbb{Z}$ is our prior knowledge of how things such as $(7 - 12) + (8 - 5)$ and $(7 - 12) \times (8 - 5)$ behave. Unofficially we can argue along the following lines:

$$(7, 12) + (8, 5) = (7 - 12) + (8 - 5) = (7 + 8) - (12 + 5) = (7 + 8, 12 + 5)$$

and

$$(7, 12) \times (8, 5) = (7 - 12) \times (8 - 5) = 7 \times 8 + 12 \times 5 - (12 \times 8 + 7 \times 5)$$
$$= (7 \times 8 + 12 \times 5, 12 \times 8 + 7 \times 5)$$

So that we could try for our official definitions:
$$(a,b) \oplus (c,d) = (a+c, b+d)$$
$$(a,b) \otimes (c,d) = (a \cdot c + b \cdot d, b \cdot c + a \cdot d)$$

Note that $\oplus$ and $\otimes$ denote new operations (on $\mathbb{Z}$) which we are trying to define by these equations. The dot and $+$ refer to ordinary multiplication and addition on $\mathbb{N}$.

Only one thing could spoil all this. We know unofficially that $(7, 12)$ is not the only ordered pair of naturals which could represent the integer $^-5$, and there is nothing special about $(7, 12)$ which gives us any reason for declaring it to be *the* representation of $^-5$ rather than $(8, 13)$, $(1, 6)$ or any of the infinitely many other possibilities. This has implications for our proposed definitions of $\oplus$ and $\otimes$, for suppose that $(a', b')$ and $(c', d')$ were different representations of the integers previously represented by $(a, b)$ and $(c, d)$. Then, for our definitions to be any use, we need some guarantee that applying them to $(a,b) \oplus (c,d)$ and $(a,b) \otimes (c,d)$ will yield the same results as $(a',b') \oplus (c',d')$ and $(a',b') \otimes (c',d')$. If we concentrate for the moment on addition, it is clear that $(a+c, b+d)$ will not be the same ordered pair as $(a'+c', b'+d')$, but this is not the problem—the guarantee we *do* require is that these two ordered pairs shall be equivalent in the sense of representing the same integer. This sense of equivalence of ordered pairs can be denoted by the symbol $\approx$ and defined in terms of $\mathbb{N}$ by: $(a,b) \approx (a',b')$ if and only if $(a+b' = b+a')$, so what we have to prove, namely

$$(a,b) \approx (a',b') \wedge (c,d) \approx (c',d') \Rightarrow ((a,b) \oplus (c,d)) \approx ((a',b') \oplus (c',d'))$$

can be expressed within $\mathbb{N}$ as

$$a'+b = a+b' \wedge c'+d = c+d' \Rightarrow a'+c'+b+d = a+c+b'+d'$$

an implication which is fairly obvious.

We have risked labouring this point, partly because students often find it difficult, and also because this same problem crops up frequently in mathematics. In essence, it always arises when we have entities which have many different representations and then attempt to define operations on them by reference to just one such representation. Only if the result of the operation is *independent* of the representation we use can we call the operation *well-defined*.

Going back to our example of addition in $\mathbb{Z}$, we can think of an integer, say $^-5$, as the whole set of ordered pairs of naturals, $\{(1,6), (2,7), (3,8), \ldots\}$, which is written as $[1, 6]$, and defined by specifying $(a,b) \in [1,6]$ if and only if $a+6 = b+1$. We then define addition on $\mathbb{Z}$ by the statement:

$$[a,b] = \{(x,y): x \in \mathbb{N}, y \in \mathbb{N}, a+y = b+x\}$$
$$\text{and } [a,b] \oplus [c,d] = [a+c, b+d]$$

Fig. 3.8

and we know that this well-defines addition because of our previous checking that if $(a',b') \in [a,b]$ and $(c',d') \in [c,d]$, then $(a'+c', b'+d') \in [a+c, b+d]$.

## 3.3 A 'DO-IT-YOURSELF' CONSTRUCTION OF ARITHMETIC BEYOND $\mathbb{N}$

With $\mathbb{Z}$ defined as in Figure 3.8, define *multiplication* on $\mathbb{Z}$ by

$$[a,b] \otimes [c,d] = [ac + bd, bc + ad]$$

**[3.5]** Check that this well-defines the operation $\otimes$ on $\mathbb{Z}$.

**[3.6]** Which set of ordered pairs of natural numbers is the zero integer? Which is the unit integer (usually called 1!)?

**[3.7]** If $z$ is the integer $[a,b]$, determine $^-z$ and $^-1$, and show that $^-1 \otimes z$ is the same as $^-z$.

**[3.8]** The ordering of $\mathbb{N}$ can be extended to $\mathbb{Z}$ by defining $[a,b] \ominus [c,d]$ if and only if $a + d < c + b$. Convince yourself that $\ominus$ is well-defined on $\mathbb{Z}$ and that it corresponds to the usual 'less than' relation.

**[3.9]** If $x, y, z$ are any *integers* and 0 is the zero integer, then show, using any facts about $\mathbb{N}$ that you require, that:

(1) One and only one of $x = y$, $x \ominus y$ or $y \ominus x$ holds.
(2) $x \oplus z \ominus y \oplus z$ if and only if $x \ominus y$.
(3) If $0 \ominus z$, $x \otimes z \ominus y \otimes z$ if and only if $x \ominus y$.
(4) If $z \ominus 0$, $x \otimes z \ominus y \otimes z$ if and only if $y \ominus x$.

### 3.3.1 The Construction of $\mathbb{Q}$ from $\mathbb{Z}$

We must first say something about notation. If $b < a$, then $a - b \in \mathbb{N}$, so we can write $[a,b]$ as $[a-b+1, 1]$ by the definition of $[a,b]$ in Figure 3.8. This in turn can be written $[n,1]$, using $n$ for the *natural number* $a - b + 1$. If $a = b$, then $[a,b] = [a,a]$, which you found was zero in exercise [**3.6**]. If $a < b$, then $b - a + 1$ is a natural number, say $m$, so $[a,b]$ can be written as $[1, m+1]$, which, as you found in exercise [**3.7**], is $-[m+1, 1]$.

Hence, the integers can be partitioned into three distinct classes:

$\{[n+1, 1]: n \in \mathbb{N}\}$—called the positive integers;
$\{[1,1]\}$—the zero integer; and
$\{[1, n+1]: n \in \mathbb{N}\}$—called the negative integers.

To minimise pedantry in notation, we shall revert to the usual $ab$ and $a + b$ and $a < b$ for multiplication, addition and 'less than' on $\mathbb{Z}$, and agree to write $n, 0$ and $-n$ for the integers $[n+1, 1], [n,n]$ and $[1, n+1]$, respectively.

Note that $2 + (-3)$ in $\mathbb{Z}$ means $[2+1, 1] \oplus [1, 3+1]$, which is $[3,1] \oplus [1,4] = [4,5] = [1,2] = [1, 1+1]$, which is $-1$.

You can easily convince yourself by doing checks like this that we have actually arrived at the familiar arithmetic for $\mathbb{Z}$.

In the remaining exercises of this section you may take for granted any of the standard properties of $\mathbb{N}$ and $\mathbb{Z}$. Since our informal understanding of a rational number is one which can be expressed as $a/b$ with $a, b$ being integers and $b \neq 0$, it is natural to define a rational number formally as a class of ordered pairs of integers. For example, 'a third' is the class consisting of $(1, 3), (11, 33), (9, 27), (-4, -12), \ldots$, which we shall write as $[1, 3]$.

Our formal definitions of $\mathbb{Q}$ and its associated addition and multiplication are displayed in Figure 3.9.

$$[a, b] = \{(x, y): x \in \mathbb{Z}, y \in \mathbb{Z}, y \neq 0, ay = bx\},$$
$\mathbb{Q}$ is the set of all such $[a, b]$,
$$[a, b] \boxplus [c, d] = [ad + bc, bd]$$
$$[a, b] \boxtimes [c, d] = [ac, bd]$$

Fig. 3.9

**[3.10]** Show that these formulae in Figure 3.9 do well-define operations on $\mathbb{Q}$.

**[3.11]** Check that they correspond to our normal understanding of addition and multiplication of fractions.

**[3.12]** Define, as ordered pairs of integers, the zero and the unity of $\mathbb{Q}$.

**[3.13]** If $r = [a, b]$, what are $-r$ and $r^{-1}$?

**[3.14]** If the ordered pair of integers $(a, b)$ has the property $ab > 0$, show that $xy > 0$ whenever $(x, y) \in [a, b]$.

**[3.15]** What familiar property of the rational number $[a, b]$ has been well-defined in the previous exercise?

**[3.16]** Let $\mathbb{Q}^+$ be the set of positive rationals. Show that $\mathbb{Q}^+$ has the following properties:

(1) If $q \in \mathbb{Q}$, one and only one of $q = 0$, $q \in \mathbb{Q}^+$ or $-q \in \mathbb{Q}^+$ holds.
(2) $\mathbb{Q}^+$ is closed under $\boxplus$ and $\boxtimes$.
(3) $q \in \mathbb{Q}^+ \Rightarrow q^{-1} \in \mathbb{Q}^+$.

**[3.17]** If $q, r \in \mathbb{Q}$, define $q \trianglelefteq r$ to mean $r \boxminus q \in \mathbb{Q}^+$. (Subtraction on $\mathbb{Q}$ is defined by $r \boxminus q = r \boxplus (-q)$). Show that $r \boxplus t \trianglelefteq s \boxplus t$ if and only if $r \trianglelefteq s$.

**[3.18]** Show that $r \trianglelefteq s \wedge s \trianglelefteq t \Rightarrow r \trianglelefteq t$.

**[3.19]** Show that $\mathbb{Q}$ is dense relative to $\trianglelefteq$—i.e. if $q \trianglelefteq r$, there exists $s \in \mathbb{Q}$ with $q \trianglelefteq s$ and $s \trianglelefteq r$. (Between any two rationals there is another.)

**[3.20]** If $[a, b]$ and $[c, d]$ are rational numbers, show that the formula $[a, b] * [c, d] = [a + c, b + d]$ *does not* well-define any operation on $\mathbb{Q}$.

## 3.4 EQUIVALENCE RELATIONS

From primary school to the frontiers of research it is no exaggeration to say that classification is the root of much mathematical activity. Shapes are classified into regular, irregular, convex, concave,...; numbers into prime, composite, odd, even, squares, cubes,.... It is often convenient for our classification to involve only non-overlapping classes (like primes and composites but unlike squares and cubes), and in this case the classification is called a partition. The following are examples of partitions.

(a) $\quad \{\ldots -8, -4, 0, 4, 8, 12, \ldots\}, \quad \{\ldots -7, -3, 1, 5, 9, \ldots\}$
$\quad \{\ldots -6, -2, 2, 6, 10, \ldots\}, \quad \{\ldots -9, -5, -1, 3, 7, \ldots\}$

is a partition of $\mathbb{Z}$ into four classes.

(b) The set of all triangles can be partitioned so that the members of each class are similar to one another but not to any triangle in a different class.

(c) The same as example (b) but with 'similar' replaced by 'congruent'.

(d) The set of all propositions using three statement variables $P$, $Q$ and $R$, and three connectives $\vee$, $\wedge$ and $\sim$, with the members of each class being truth-functionally equivalent to one another but not to any proposition outside the class.

(e) The set of all lines in the plane partitioned so that two lines are in the same class if and only if they are parallel.

(f) $\mathbb{N}$, partitioned into classes, $\{x : 1 \leqslant x < 10\}$, $\{x : 10 \leqslant x < 10^2\}$, $\{x : 10^2 \leqslant x < 10^3\}, \ldots$.

(g) The points $(x, y)$ in the plane partitioned so that $(x_1, y_1)$ and $(x_2, y_2)$ are in the same class if and only if $x_1^2 - y_2^2 = x_2^2 - y_1^2$.

There are several points to make about these examples.

(1) In the first six examples it is quite clear that we have a genuine partition—the classes don't overlap and between them they cover the whole set in question. In the last example it is not clear that we have defined a partition of the plane. What we have done is defined a relationship between two points and said 'Put these points in the same class if the relationship holds between them, and not otherwise'. But what if the relationship holds between points $P$ and $Q$, and between $Q$ and $R$, but not between $P$ and $R$? Our instruction would then turn out to be impossible to obey. Why? [**3.21**]

(2) In fact, the situation envisaged above doesn't happen in example (g), and you should check that this is so. [**3.22**]

By way of contrast, consider the following rather fanciful example: I attempt to partition the set of all cups of coffee by saying that two cups go into the same class if and only if I can't taste any difference between them—and we keep things simple by assuming that I can only taste the difference if the sugar contents differ by at least five grains. Why will this not work? [**3.23**]

(3) In many of the examples the members of the same set of the partition have some obvious feature in common: in (a) it is the remainder on division by 4; in (b) the members of the same set have the same shape (and we ignore differences in size and position); in (e) direction is the important property; in (f) it is the number of digits; and so on.

(4) There can be infinitely many classes in the partition, each containing infinitely many members, as in example (e). There can be infinitely many classes but all of finite size, as in (f). There may be only finitely many classes each containing infinitely many members as in (a),... or we can have any variation of these. How many classes are there in (d)? [**3.24**]

(5) In some examples such as (a) and (f) the partition is specified explicitly. In the others it is specified indirectly by giving the necessary and sufficient condition for two elements to belong to the same class of the partition. In some it is easy to change from one to the other by a simple rephrasing. For example, (a) can be expressed as '$x, y$ belong to the same class if and only if $x - y$ is divisible by 4', and in (e) each class could be specified by giving the common gradient of its member lines. Sometimes, as in (g), the rephrasing is not so easy and we need some way of telling whether a given relation corresponds to a partition or not. To tackle this problem, it is first necessary to say exactly what we mean by a relation.

Suppose that our set is $\{1, 2, 3, 4\}$ and we say '$x$ is greater than $y$'. The ordered pairs $(x, y)$ from $\{1, 2, 3, 4\}$ for which this relationship holds are $(4, 1)$, $(4, 2), (4, 3), (3, 1), (3, 2), (2, 1)$ and no others. This, formally, is what a relation on a set $S$ is—just a subset of $S \times S$, although usually there is some ordinary phrase which captures the idea behind the relation, such as 'is greater than', 'is a factor of', 'is perpendicular to', 'has the same set of prime factors as'..., and so on.

We now define three important properties which some relations have. We use $\mathcal{R}$ for the relation and $S$ for the set it lives on, and we write $x\mathcal{R}y$ to express the fact that $x$ has the relation $\mathcal{R}$ to $y$ (or $(x, y)$ is a member of $\mathcal{R}$).

$\mathcal{R}$ is *reflexive* if $x\mathcal{R}x$ for all $x$ in $S$.
$\mathcal{R}$ is *symmetric* if $y\mathcal{R}x$ whenever $x\mathcal{R}y$.
$\mathcal{R}$ is *transitive* if $x\mathcal{R}z$ whenever $x\mathcal{R}y$ and $y\mathcal{R}z$.

The following examples should give you a feel for what is involved.

(a) On $\mathbb{N}$ the relation 'is a factor of' is reflexive and transitive but not symmetric.

(b) 'is parallel to', on any set of lines, has all three properties.

(c) $>$ on $\mathbb{N}$ is transitive but neither symmetric nor reflexive.

(d) 'differs by at least 2 from' on $\mathbb{N}$ is symmetric but not reflexive or transitive.

(e) On the set $\{3, 5, 7, 22\}$ let $\mathcal{R}$ be defined by specifying that $x\mathcal{R}y$ if and only if $x$ differs from $y$ by 2 or 4, or $x$ is a prime factor of $y$. This rather contrived example is symmetric and transitive but not reflexive.

Before going further, it would be wise to satisfy yourself that these examples have the properties claimed for them. [**3.25**]

Whether a relation has these properties or not depends just as much on the set $S$ as on the 'description' of $\mathcal{R}$. For example, had $\mathcal{R}$ in example (e) been defined on $\{3, 5, 7\}$ instead of $\{3, 5, 7, 22\}$, it would have been reflexive.

Similarly, 'is a sister of' is a symmetric relation on the set of all female people, but not on the whole human race.

We have seen in example (e) that it is possible for a relation to be both symmetric and transitive without being reflexive, so the following 'proof' that symmetry and transitivity together imply reflexivity must be bogus.

Let $\mathcal{R}$ be a relation on $S$ which is symmetric and transitive. We 'prove' that $a\mathcal{R}a$ for all $a$ in $S$. Choose any $b$ in $S$ for which $a\mathcal{R}b$. Then by symmetry $b\mathcal{R}a$. Now that we have $a\mathcal{R}b$ and $b\mathcal{R}a$, transitivity ensures that $a\mathcal{R}a$.

Where is the fallacy? [**3.26**] [*Hint*: The word 'whenever' in the definition of symmetry is vitally important.]

If $\mathcal{R}$ is a relation on $S$ and $x \in S$, we use $C_x$ to denote the subset of elements $y$ of $S$ for which $x\mathcal{R}y$. Formally, $C_x = \{y : y \in S \wedge x\mathcal{R}y\}$. The following three examples lead us back to partitions.

(a) On $\mathbb{N}$, $x\mathcal{R}y$ means that $x$ and $y$ differ by at most 5. Check that $C_2$ and $C_{10}$ are different sets, but that they intersect, so that the distinct $C_x$'s do not partition $\mathbb{N}$. [**3.27**]

(b) On $\mathbb{N}$, $x\mathcal{R}y$ means $x$ is a prime factor of $y$. Here the situation is worse. Not only do distinct classes overlap (6 is in $C_2$ and $C_3$ and $C_2 \neq C_3$), but the number 1 is not in any class, so we fail on two counts to get a partition of $\mathbb{N}$.

(c) Our final example before proving the basic theorem of this section goes back to the beginning of Section 3.4: $\mathcal{R}$ is a relation on $\mathbb{Z}$ defined by $x\mathcal{R}y$ if and only if $x - y$ is divisible by 4. $C_0, C_1, C_2$ and $C_3$ are clearly the sets written out in that example, and they partition $\mathbb{Z}$. Any other class $C_n$ is identical with one of these four (e.g. $C_{79} = C_3, C_{-10} = C_2$), so the *distinct* classes do partition $\mathbb{Z}$. Note that this $\mathcal{R}$ is reflexive, symmetric and transitive. Any such relation is called an *equivalence relation*, and its classes are called *equivalence classes*, which brings us to the following very important theorem.

---

**Fundamental Theorem on Equivalence Relations**

(i) If $\mathcal{R}$ is an equivalence relation on a set $S$, then the distinct equivalence classes partition $S$. (ii) Conversely, if any partition of $S$ is given, then the subsets of $S$ which make up the partition are equivalence classes of some equivalence relation.

---

*Proof*

(i) Every element $x$ of $S$ is in some equivalence class, namely $C_x$, because, by the reflexive property of $\mathcal{R}$, $x\mathcal{R}x$. This establishes that the union of all the classes is the whole of $S$. It remains to show that any two classes $C_x$ and $C_y$ are either disjoint or identical, so that the *distinct* classes do form a partition of $S$. To this end, suppose that $C_x$ and $C_y$ do have some element in common, say $z$, and let $w$ be an arbitrary member of $C_x$. Then $x\mathcal{R}w$ and $x\mathcal{R}z$, since $w$ and $z$ are in $C_x$; and $y\mathcal{R}z$, since $z$ is in $C_y$.

So, using the symmetry of $\mathcal{R}$, we have $y\mathcal{R}z$, $z\mathcal{R}x$ and $x\mathcal{R}w$. Applying the transitive property to the first two of these, we have $y\mathcal{R}x$, and, combining this with $x\mathcal{R}w$ and applying transitivity again, we get $y\mathcal{R}w$. That is, $w \in C_y$. So we have proved that each element in $C_x$ is also in $C_y$; hence, $C_x \subseteq C_y$.

The reverse inclusion is obtained by copying this proof with the roles of $x$ and $y$ interchanged. Hence, $C_x = C_y$. This was proved just on the assumption that $C_x$ and $C_y$ had an element in common, so what we have really proved is that any two classes $C_x$ and $C_y$ either have all their elements in common ($C_x = C_y$) or no elements in common ($C_x \cap C_y = \emptyset$), which is what was required.

(ii) The converse is rather trivial: If we have a set of subsets of $S$ which partition $S$, we can define a relation $\mathcal{R}$ on $S$ by saying $x\mathcal{R}y$ if and only if $x$ and $y$ are in the same subset of the partition. The verification that this makes $\mathcal{R}$ an equivalence relation is easy [**3.28**]. □

We conclude this chapter with some exercises on equivalence relations.

[**3.29**] Verify that the relation $\approx$ defined on $\mathbb{N} \times \mathbb{N}$ by $(a,b) \approx (c,d)$ if and only if $a + d = b + c$ is an equivalence relation.

[**3.30**] Do the same for $(a,b)\mathcal{R}(c,d)$ if and only if $ad = bc$, where $\mathcal{R}$ is a relation on $\mathbb{Z} \times (\mathbb{Z}\setminus\{0\})$.

Readers of Sections 3.2 and 3.3 should recognise the equivalence classes involved in the two examples above!

[**3.31**] How many different equivalence relations are there on a three-element set?

[**3.32**] A relation $\mathcal{R}$ on $\{a,b,c,d\}$ consists of the ordered pairs $(a,b)$, $(a,c)$, $(a,a)$, $(b,d)$, $(c,c)$. Find minimal sets of pairs to add to these in order to make $\mathcal{R}$ (i) reflexive, (ii) symmetric, (iii) transitive. Then do the same for the set $\{(a,b),(a,c),(a,a),(c,c)\}$.

[**3.33**] Let $n$ be a fixed natural number and $\alpha$ be a relation defined on $\mathbb{N}$ by $a\alpha b$ if and only if $a^2 - b^2$ is a multiple of $n$. Show that $\alpha$ is an equivalence relation and work out the equivalence classes in each of the three cases $n = 7$, $n = 8$ and $n = 10$.

[**3.34**] Let $T$ be the set of integers between 2 and 100 inclusive. $\mathcal{R}$ is a relation defined on $T$ as follows: $a\mathcal{R}b$ if and only if, in the factorisations of $a$ and $b$ into primes, their lowest prime factors are raised to the same power. (For example, $12\mathcal{R}45$ because $12 = 2^2 \times 3^1$ and $45 = 3^2 \times 5^1$, but $\sim 35\mathcal{R}63$ because $35 = 5^1 \times 7^1$ and $63 = 3^2 \times 7^1$.) Show that $\mathcal{R}$ is an equivalence relation and find all the equivalence classes.

[**3.35**] Show that the relation $\tau$ defined on all points of the plane by $(a,b)\tau(c,d)$ if and only if $c^2b = a^2d$ is *not* an equivalence relation, but that if we restrict $\tau$ to all points of the plane except those on the axes, it *is* an equivalence relation. In this case describe the equivalence classes.

[**3.36**] Show that a relation $\rho$ on a set $A$ is an equivalence relation if and only if it satisfies $a\rho b \wedge b\rho c \Rightarrow c\rho a$ for all $a$, $b$, $c$ in $A$, and is reflexive.

[**3.37**] $S$ is an infinite set and $\Omega$ is the set of all infinite subsets of $S$. $\mathscr{R}$ is a relation defined on $\Omega$ as follows: $A \mathscr{R} B$ means that $A \setminus B$ and $B \setminus A$ are both finite. Prove that $\mathscr{R}$ is an equivalence relation. Taking $S$ to be $\mathbb{N}$, find (i) two subsets in the same equivalence class and (ii) two subsets in different classes.

# 4 The Real Numbers—in which we find holes in the number line and pay the price for repairs

> 'The paradox is now fully established that the utmost abstractions are the true weapons with which to control our thoughts of concrete fact.'
>
> Alfred North Whitehead

## 4.1 HOLES ARE SUSPECTED

As advertised in the previous chapter, we shall take for granted the elementary arithmetic of fractions and all the generally illogical manoeuvrings it takes to get there. What we now need is a working definition of the set $\mathbb{Q}$ of rational numbers and a specific interpretation or model of them. Both are easy: the model is the familiar number line in which the rational number $x$ is represented as a distance $x$ along the line from 0, to the left or right, depending on whether $x$ is negative or positive; and our (semi-formal) definition of a rational number is any number which can be expressed as the ratio between two integers, $n/m$, with the proviso that $m$ is not zero.

We choose to make a fuss over the step from $\mathbb{Q}$ to $\mathbb{R}$ rather than any of the other steps, because it is the only step which contains a real surprise. It is the step at which our faith that the geometry of the line and the arithmetic of numbers are essentially the same begins to show cracks. If we are thinking of numbers as markers for the points making up the line, it is not immediately obvious that we need any numbers other than the rationals. However, we shall see shortly that in order to restore the belief that our knowledge of the line can all be expressed in terms of numbers, we have to *invent* an extension of $\mathbb{Q}$. Let us begin to see why this is so.

First, it seems reasonable to suppose that once we have chosen our unit of length (say a metre), then any length should be expressible as a whole number of $n$ths of a metre, provided that we are prepared to make $n$ sufficiently big. An alternative way of expressing this basic intuition is in terms of the density of the rational numbers. Suppose that we take a point on the line a distance $l$ to the right of 0, and let $a$ and $b$ be two rationals with $a < l < b$. Now, however close together $a$ and $b$ may be, there will be infinitely many more rationals between them, so surely one of them must coincide with $l$. Sadly, this intuition, though reasonable, is wrong. With hindsight we can see why the argument must fail, because the claim is essentially that if a set of points (say those between 2 and 3) is everywhere infinitely closely packed, then it must contain *every* point of the interval. That this is false can be seen by generating such a set by successive bisection:

$$\{2, 3, 2\tfrac{1}{2}, 2\tfrac{1}{4}, 2\tfrac{3}{4}, 2\tfrac{1}{8}, 2\tfrac{3}{8}, 2\tfrac{5}{8}, 2\tfrac{7}{8}, 2\tfrac{1}{16}, \ldots\}$$

This set is clearly infinitely closely packed in every bit of the interval, but it can't possibly contain the point $2\tfrac{1}{3}$—or, indeed, any rational whose denominator (in lowest terms) is not a power of 2. The real surprise is that, if we try to patch up the argument by making no restriction on the denominators, if we throw in *all* the rationals between 2 and 3, we still leave out some points. This is what is hard to swallow, and no one should expect you to until an example of one of these 'left-out' points has been produced. To do just that, consider the length we usually denote by $\sqrt{5}$.

## 4.2 HOLES ARE FOUND

We can't explain the symbol $\sqrt{5}$ by defining it as 'that number whose square is 5', because the only numbers whose existence we officially acknowledge so far are the rationals, and we are aiming to show that $\sqrt{5}$ is not one of them! However, we can show that $\sqrt{5}$ can be reasonably interpreted as a length (and, hence, as a point on the line) in two ways. First, if we accept Pythagoras' theorem, $\sqrt{5}$ is just the length of the diagonal of a $2 \times 1$ rectangle. Second, we may choose to remain sceptical about Pythagoras and rely instead on our feeling for geometric continuity: Imagine a square expanding from $2 \times 2$ to $3 \times 3$ continuously. We expect its area to change continuously from 4 to 9, so at some time its area must be 5 and at this instant its side length can only be described as $\sqrt{5}$! So $\sqrt{5}$ is respectable as a length; why is it not rational? Let us suppose that it is, and explore the consequences of this supposition. (This is a classic example of proof by contradiction.)

If $\sqrt{5}$ is a rational, it is $a/b$, where $a$ and $b$ are natural numbers. Hence, $a^2 = 5b^2$.

Now consider factorisations of $a$ and $b$ into primes, and let $m$ and $n$ be the number of times that the factor 5 occurs in $a$ and $b$, respectively. By squaring each factor, we get prime factorisations of $a^2$ and $b^2$ which contain 5 $2m$ and $2n$ times, respectively. Finally, by inserting an extra 5 in the factorisation of $b^2$, we get a factorisation of $5b^2$ which contains $2n + 1$ 5s. So we have prime factorisations of the same number ($a^2 = 5b^2$) one of which contains an even number ($2m$) of 5s and the other an odd number ($2n + 1$). This is impossible by UPF (Chapter 2).

To recap, we have made a supposition, one of the consequences of which was the denial of something we know to be true. Hence, that supposition must be wrong. That is, $\sqrt{5}$ cannot be rational. For an alternative proof that $\sqrt{5}$ is irrational, see Section 7.1.

So, by demanding that numbers correspond exactly to our geometric intuitions about lengths, areas and continuous change, there is no escape from accepting $\sqrt{5}$ as a fully respectable 'number'. Much of this chapter will be taken up with giving $\sqrt{5}$ a still greater appearance of respectability by attempting to define it (and irrational numbers in general) in terms of rational numbers, just as the rationals were defined in terms of integers and integers in terms of naturals in Chapter 3. By doing this, we aim to make the notion of mathematical existence more precise, but first it is worth thinking about the sense in which irrational numbers definitely *do not* exist. Let us provisionally accept the evidence (or faith) of modern physics that the universe is constructed entirely from a small number of types of elementary particles with fixed masses, $m_1, m_2, \ldots, m_n$. For simplicity assume for the moment that there is just one type, with mass $m$. Then it would be impossible (in principle as well as in practice) to split any real object or collection of objects into two bits whose ratio of masses was $\sqrt{5}:1$, and $\sqrt{5}$ could only exist as a mental fiction if we decided to commit ourselves to mass ratios as the sole interpretation of numbers. Even in the general case of several different types of fundamental particles, some irrational numbers may exist as mass ratios but infinitely many wouldn't! To give a specific example, suppose that there were two fundamental particles of masses 1 and $\sqrt{5}$. Show that no two collections of particles could have masses in the ratio $\sqrt{7}:1$. [**4.1**] Chapter 8 has some exercises in a similar vein.

## 4.3 HOLES ARE IMPORTANT

Historically the distinction between lengths and numbers is important, as can be seen from the following famous Greek proof that equiangular triangles have their corresponding side lengths in proportion.

Let the triangles ABC and PQR of Figure 4.1 be the given equiangular triangles, with corresponding equal angles marked as in the diagram. If we now superimpose the triangles as in Figure 4.2 and assume that the ratio of lengths PQ:AB is a rational number $n/m$, $n$ and $m$ being natural numbers with $m \leq n$, we can divide PQ into $n$ equal parts, draw lines through the points of subdivision all parallel to QR, and BC will be one of these lines (the $m$th one counting from A).

If we now draw lines parallel to PQ, through all the points of subdivision of PR, the small triangles generated are all congruent. [**4.2**] Hence, the segments into which PR is split are all the same length, there are $n$ of them and $m$ of them make up AC. So PR:AC = $n/m$.

An identical proof that QR:BC = $n/m$ can be derived by superimposing B on Q. [**4.3**]

Clearly, this proof only works because PQ/AB is rational, because otherwise we could not carry out the construction on which the proof depends. Another way of saying that PQ/AB is rational is that there exists a length

# THE REAL NUMBERS

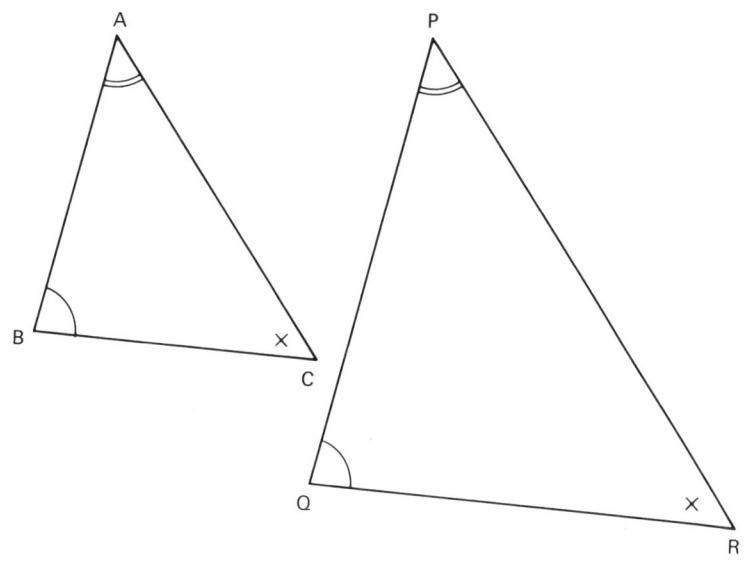

**Fig. 4.1**

unit (in this case one *m*th of AB) in terms of which both AB and PQ are a whole number of units. The Greeks called such lengths commensurable, in contrast to pairs of lengths such as $\sqrt{5}$ and 1, which we have proved are incommensurable.

Our theorem above said that if two triangles are equiangular, then they

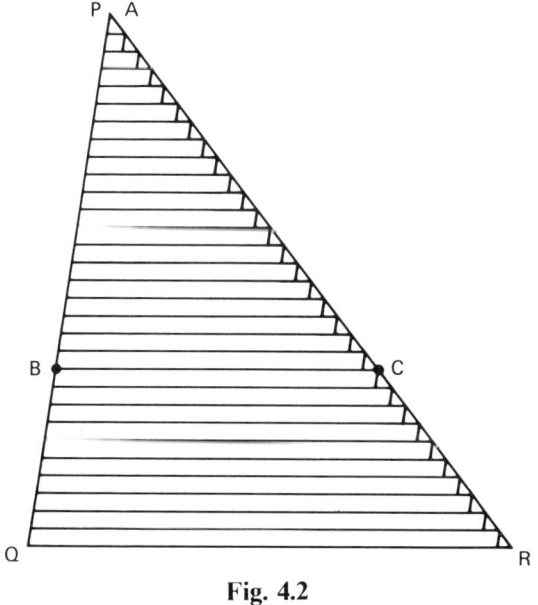

**Fig. 4.2**

are similar. We may suspect that it is true in general, even though we have only proved it for the case when a pair of corresponding sides are commensurable, and set about finding a proof which works for the incommensurable case too. However, the issue goes much deeper than this: we cannot even *state* the theorem legitimately until we have a proper theory of the arithmetic of irrational numbers. The reason for this alarming state of affairs is that the notion of similarity involves ratios of side lengths—in other words, division of numbers—and we haven't even managed to *define* irrational numbers yet, let alone divide one by another!

We begin to tackle this problem in Section 4.6, and you will lose nothing by jumping straight to this section now. Meanwhile we pause to give a purely geometric method of finding the greatest common measure of two commensurable line segments, and then, in Section 4.5, make an interesting link with this and with Chapter 2 by discussing a method of finding highest common factors without doing any factorising!

## 4.4 HOW TO MEASURE A LINE

Suppose that we are given two straight line segments AB and CD with AB longer than CD, and suppose that another line segment PQ is a common measure of them. By this we mean that AB and CD are both equal in length to a whole number of PQs placed end to end. We express this by saying that PQ 'measures' AB and CD. For example, if AB is 108 cm and CD is 93.6 cm, then 1 mm and 2 mm are obvious common measures. Another one, not so obvious, is 72 mm.

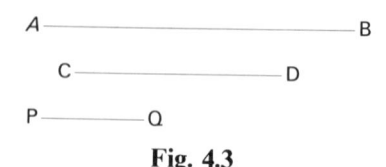

```
A————————————————B
   C——————————————D
   P————Q
```

**Fig. 4.3**

Tweedledum now proposes the following problem to Alice.

TWEEDLEDUM: Here are two lines AB and CD. Find a common measure for them.
ALICE: How do I know they *have* a common measure? They may have lengths 1 and $\sqrt{5}$.
TWEEDLEDUM: Sorry, I forgot to tell you. They do actually have a common measure; I'll give you that information for free.
ALICE: Right, in that case I'll just fetch my ruler, measure the lines, then a bit of arithmetic should do the trick.
TWEEDLEDUM: Sorry, that's not in the rules. No rulers allowed.
ALICE: If you're going to make the rules up as you go along, this game isn't going to be worth playing.

TWEEDLEDUM: A fair point. OK, I'll try to be quite explicit about the rules. You are allowed a pencil and a pair of compasses but no ruler, so you have to *construct* a line which measures AB and CD, not just say how long it would be. So you have to do this in the spirit of Greek geometry—pure construction, no measuring and no arithmetising.

ALICE: Well, I'll have a go, but what if the common measure turns out to be too small for this pencil and this pair of compasses to cope with?

TWEEDLEDUM: Remember, I did say *in the spirit* of Greek geometry, not the practicalities of it. So you have the power of Greek gods—that is, all powerful in some respects but pretty helpless in others. We'll assume you have perfect eyesight so that you can distinguish between your compass pencil drawing an arc through a point or just missing it, however minute the miss may be. And your compasses are perfect in the sense of being able to draw an arc of any radius, however big or small, with perfect accuracy. But your powers definitely don't extend to having access to a ruler. Now, since you have forced me to go into this detail, let's make the problem even more specific and ask not just for some common measure but for the longest.

Having set up the problem in classical Greek terms, we could now go on to specify the construction which will produce the longest common measure of AB and CD. However, to present the method without explaining why it works and without digging a little deeper into the explanation would be totally against the philosophy behind this book—which is our excuse for the following digressions, which are carried further in the exercises.

**Digression 1** When a set (of numbers or lengths) has infinitely many members, then, even if its members are all less than some fixed number, there is no guarantee that the set has a biggest member. For example, the set $M = \{\frac{1}{2}, \frac{3}{4}, \frac{7}{8}, \frac{15}{16} \ldots\}$ has no biggest member. Why not? [**4.4**] Now the set of common measures of AB and CD has at least one member—remember that this was the given information. Deduce from this that it has infinitely many members [**4.5**], and then convince yourself that although this set, like $M$, is infinite and bounded above (there is no member longer than CD), it must have, unlike $M$, a longest member. [**4.6**]

**Digression 2** One of the pleasures of mathematics is finding connections between ideas where no such connection was previously suspected. As an example of this consider the case of two sets which may have cropped up in quite different contexts and which are therefore defined or described in quite different ways, but which turn out to be identical. Set $A$ is just all the multiples of 6, including the negative ones, so $A = \{\ldots, -24, -18, -12, -6, 0, 6, 12, 18, \ldots\}$ and $B$ is the set of all integers expressible as the sum of a multiple of 30 and a multiple of 18, so $B = \{30a + 18b : a \in \mathbb{Z}, b \in \mathbb{Z}\}$.

Now it is easy to see that every member of $B$ is divisible by 6, so is a member of $A$. This establishes that $B \subseteq A$. If we could also establish that

$A \subseteq B$, this would complete the demonstration that $A = B$ (by the principle mentioned in Section 1.3). To do this we only need to show that 6 is in $B$, because this would mean that $6 = 30i + 18j$ for some integers $i$ and $j$, and then any multiple of 6, say $6k$, could be written as $30(ik) + 18(jk)$, showing that it, too, was in $B$. A bit of intelligent trial and error soon shows that $i = -1, j = 2$ gives us what we want, and the proof is complete.

□

Two more examples of this technique are given in the end-of-chapter exercises. Our construction of the longest common measure of AB and CD depends on showing that the set of common measures of one pair of lines is identical with the set of common measures of another pair. This is the point of our final diversion, which will bring us back on course for doing the construction to solve Tweedledum's problem.

**Digression 3** Let $M$ be the set of common measures of lines of lengths $x$ and $y(x > y)$, and let $M'$ be the set of common measures of lines of lengths $y$ and $x - y$. Then $M = M'$.

*Proof* Let $l$ be the length of any common measure of the first pair of lines. Then $x = ln_1$ and $y = ln_2$ for some natural numbers $n_1$ and $n_2$. $l$ is also a common measure of the second pair, because $y = ln_2$ and $x - y = l(n_1 - n_2)$. This shows that $M \subseteq M'$. Prove in the same way that $M' \subseteq M$ [**4.7**], to establish that $M = M'$.

□

Now for our construction: we have illustrated the first few steps in Figure 4.4.

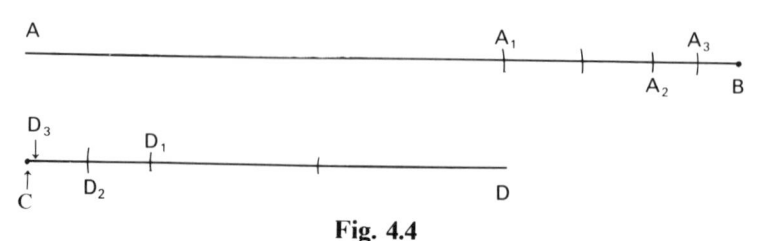

**Fig. 4.4**

With centre A and radius CD, draw an arc to cut AB. In general, we would continue this process by taking centre at this first cut, and the same radius and cutting AB again, and so on. In our illustration only one application of this process is possible (the next one would overshoot B). In any case, we stop when we have reached the last cut point which doesn't overshoot B, and call this $A_1$.

The length of $A_1B$ is AB − some multiple of CD, so, by the sort of argument used in Digression 3, AB and CD have the same set of common measures as CD and $A_1B$.

Now repeat the process on $A_1B$ and CD. With centre D and radius $BA_1$, cut DC. Repeat with centre on the cut (if possible),... and so on, calling the last cut point $D_1$. Now $CD_1$ and $A_1B$ have the same set of common measures.

Repeat with radius $CD_1$ to cut off chunks from $A_1B$ to arrive at $A_2$. $A_2B$ and $CD_1$ also have the same set of common measures as the previous two pairs.

... and so on.

The pairs of line segments we produce by this process are:

$$\frac{AB}{CD}, \frac{CD}{A_1B}, \frac{A_1B}{CD_1}, \frac{CD_1}{A_2B}, \frac{A_2B}{CD_2}, \frac{CD_2}{A_3B}, \frac{A_3B}{CD_3}, \ldots$$

and they all have the same set of common measures. In particular (and appealing to our first digression), it follows that all the pairs have the same *longest* common measure.

By the way the successive segments were constructed we have:

$$AB \geqslant 2A_1B, \quad A_1B \geqslant 2A_2B, \quad A_2B \geqslant 2A_3B, \ldots$$

and

$$CD \geqslant 2CD_1, \quad CD_1 \geqslant 2CD_2, \quad CD_2 \geqslant 2CD_3, \ldots$$

and from these strings of inequalities it is clear that

$$AB \geqslant 2A_1B \geqslant 4A_2B \geqslant 8A_3B \geqslant \ldots \quad \text{and} \quad CD \geqslant 2CD_1 \geqslant 4CD_2 \geqslant 8CD_3 \geqslant \ldots$$

so

$$A_iB \leqslant \frac{AB}{2^i} \quad \text{and} \quad CD_{i-1} \leqslant \frac{CD}{2^{i-1}}$$

Hence, $A_iB$ and $CD_{i-1}$ can be made as small as we like by taking $i$ sufficiently big—that is, taking the construction through sufficiently many steps. In particular, if $m$ is the length of the longest common measure of AB and CD, then some $A_iB$ and $CD_{i-1}$ will be shorter than $m$. But this is impossible, because $A_iB$ and $CD_{i-1}$ also have $m$ as their longest common measure.

The only way to escape this contradiction is to see that it only arises if we *can* carry on the construction for as many steps as we like, and the only thing which could prevent this is that either some $A_i$ coincides with B or some $D_i$ coincides with C.

By looking again at the specification of the construction we see that (a) if $D_i$ coincides with C, $D_{i-1}C$ is a whole number of $BA_i$s, and (b) if $A_i$ coincides with B, $A_{i-1}B$ is a whole number of $CD_{i-1}$s.

If (a) happens, then $D_{i-1}C$ and $BA_i$ have $BA_i$ as their longest common measure, and if (b) happens, $A_{i-1}B$ and $CD_{i-1}$ have $CD_{i-1}$ as their longest common measure. In either case the construction has produced a line segment ($BA_i$ or $CD_{i-1}$) which is the longest common measure of the original segments AB and CD.

Putting our result arithmetically: if AB/CD is rational, this construction will terminate in a finite number of steps; if not, it won't.

At each step we are, in a sense, 'dividing' $A_iB$ by $D_iC$ and getting a remainder of $A_{i+1}B$. We pursue this analogy in the next section, to obtain an arithmetic process for finding the highest common factor of two natural numbers, without doing any factorising—as promised in Section 4.3 and in Chapter 2.

## 4.5 EUCLID'S ALGORITHM

Suppose that we wish to find the highest common factor of 5767 and 4453, which we shall abbreviate as hcf(5767, 4453). This is rather like finding the longest common measure of two lines with these lengths, so let us copy what we did in the previous section.

First, we chop off 4453 as many times as we can (one in this case) from 5767 and see what we have left:

$$5767 \div 4453 = 1, \text{ with a remainder of } 1314$$

Next chop off 1314 as many times as we can (three in this case) from 4453 and see what is left:

$$4453 \div 1314 = 3, \text{ with a remainder of } 511$$

At this stage we leave you to prove in exactly the same way as in Section 4.4 that $\text{hcf}(a, b) = \text{hcf}(b, a - kb)$ for any integers $a, b, k$. [**4.8**]

Applying this to the two steps we have carried out so far,

$$\text{hcf}(5767, 4453) = \text{hcf}(4453, 1314) = \text{hcf}(1314, 511)$$

Continuing the process as far as we can, and writing it out systematically, we get

$$\begin{aligned}
5767 &= 1 \times 4453 + 1314 \\
4453 &= 3 \times 1314 + 511 \\
1314 &= 2 \times 511 + 292 \\
511 &= 1 \times 292 + 219 \\
292 &= 1 \times 219 + 73 \quad * \\
219 &= 3 \times 73 + 0
\end{aligned}$$

and from the line marked * we obtain

$$\text{hcf}(5767, 4453) = \text{hcf}(219, 73)$$

which, from the next line, we see is 73.

This process is called Euclid's algorithm for finding the hcf of any two natural numbers. It always works in the sense of terminating with a zero remainder after a finite number of steps. Why? [**4.9**] The required hcf is the last non-zero remainder.

We suggest that you try the algorithm out on a few pairs of numbers, and think about how you could organise the calculations as efficiently as possible on a calculator [**4.10**]—one with two memories is particularly convenient—or on a home computer.

# THE REAL NUMBERS

Euclid's algorithm is not just a pleasant curiosity. It is a starting point for many important topics in number theory, such as continued fractions and Diophantine equations, and it provides a method of simplifying significantly our proof of UPF given in Chapter 2. Tempting as it would be to digress to these topics now, it would lead us away from the main business of this chapter, which is to explore some of the issues involved in getting from $\mathbb{Q}$ to $\mathbb{R}$, so we resist the temptation. [One of the best 'do it yourself' investigational approaches to number theory is *A Pathway into Number Theory*, by R. P. Burn.]

## 4.6 SOME ROUTES FROM $\mathbb{Q}$ TO $\mathbb{R}$

In Chapter 3 we outlined (or rather, you did if you did the exercises!) a systematic procedure for building up the number system by defining new objects in terms of familiar ones. For example, $\mathbb{Z}$ could be identified as a set of equivalence classes of ordered pairs from $\mathbb{N}$. We continue this process now by exploring three ways in which the set of *all* points on the line—that is, the set $\mathbb{R}$ of real numbers—can be grasped arithmetically in terms of $\mathbb{Q}$.

### 4.6.1 Route 1: Nested Intervals

First, it is useful to introduce some standard shorthand: a *closed interval I* denoted by $[a, b]$ means the set of all numbers between $a$ and $b$, including $a$ and $b$ themselves. In set notation this can be written as $\{x : a \leq x \leq b\}$. The corresponding set obtained by excluding the end-points is called an open interval and is denoted by $(a, b) = \{x : a < x < b\}$.

Figure 4.5 represents a sequence of closed intervals $I_1, I_2, \ldots$ in which $I_i = [p_i, q_i]$. Since our official knowledge extends only to $\mathbb{Q}$, we take the end-points of each interval to be rational. The sequence is said to be nested, which means that each interval is a subset of the set of points in the preceding interval. Numerically this is equivalent to the strings of inequalities: $p_1 \leq p_2 \leq p_3 \leq \ldots$, and $q_1 \geq q_2 \geq q_3 \geq \ldots$, and, for each $i$ and $j$, $p_i \leq q_j$.

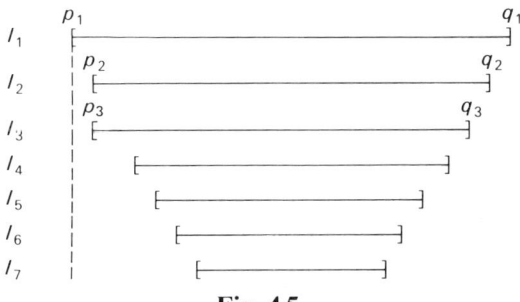

**Fig. 4.5**

Our geometric intuition can be nudged towards telling us that there will be some points (perhaps only one) in all the intervals of the sequence by the following 'thought experiment'. Note that the set of points which belong to all of the first $i$ intervals is just the set of points in the $i$th interval, because of the nesting property. So the 'size' of the set of points in all the intervals doesn't seem to depend on where the intervals live on the number line, provided that they are nested. So let us mentally shift them all to the left so that all their left-hand end-points coincide at $p_1$. This has been done in Figure 4.6. Now it is clear that at least one point, $p_1$, is in all the intervals $I_1^1, I_2^1, I_3^1, \ldots$, so when we translate them back to their original positions, this property of having at least one point in common should be preserved.

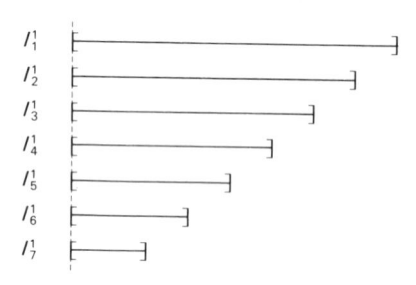

**Fig. 4.6**

Having made it reasonable to believe that nested closed intervals have at least one point in common, let us now restrict consideration to nested sequences whose intervals become arbitrarily small. In this case it is virtually obvious that the sequence has only one point common to all its intervals. To see this, argue by contradiction and suppose that $x_1$ and $x_2$ are both common points (and suppose that $x_1$ is the smaller). Then all the points between these two are also common points, so we have an interval $[x_1, x_2]$ which is included in all the $I_i$'s. But these have lengths approaching zero, so parts of $[x_1, x_2]$ must lie outside some $I_j$, and we have our contradiction.

From now on we shall use the term *nest* to denote a sequence of nested closed intervals whose lengths approach zero and whose end-points are rational numbers. In the step from $\mathbb{Q}$ to $\mathbb{R}$ the nests will play the same role as the ordered pairs of naturals in the step from $\mathbb{N}$ to $\mathbb{Z}$. To be precise, we identify a point (a real number) with the set of all nests having that point as its intersection. To avoid circularity in the definition, we have to be able to say what we mean by two nests representing the same real number *in terms of rational numbers only*. This is not as tricky as it may sound. Note that any two closed intervals with rational end-points can only intersect in one of three ways: the intersection is empty; it is a single point; or it is another interval of the same type. This is illustrated in Figure 4.7.

So, if two nests, $I_1, I_2, \ldots$ and $J_1, J_2, \ldots$ are to converge to the same point, their sequence of intersections, $I_1 \cap J_1, I_2 \cap J, \cdots$ must all be non-empty. You should check that in this case $I_1 \cap J_1, I_2 \cap J_2, \cdots$ is also a nest. [**4.11**]

**Fig. 4.7**

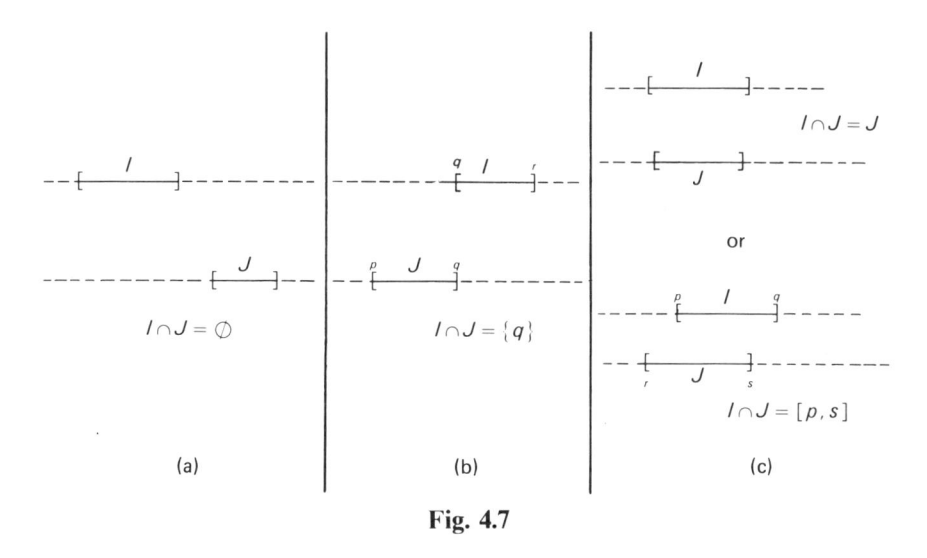

Conversely, if all the intersections are non-empty, the original two nests converge to the same point. [4.12] This condition on the intersections will be our criterion for the equivalence of two nests, and we have managed to state it with reference to $\mathbb{Q}$ only.

> **Definition**  Nests $I_1, I_2, I_3, \ldots$ and $J_1, J_2, J_3, \ldots$ are *equivalent* if the intersections $I_1 \cap J_1, I_2 \cap J_2, \cdots$ are all non-empty.

The next steps, which we defer until Chapter 7, are to say what it means to add, subtract, multiply and divide two nests, to solve the corresponding 'well-defined' problem so that these operations do actually define addition, subtraction, multiplication and division of real numbers, and to check that we can prove that this abstract system has all the properties we commonly take for granted in $\mathbb{R}$.

### 4.6.2 An Optional Philosophical Digression

Before discussing other methods of grasping those elusive members of $\mathbb{R}$ which are not in $\mathbb{Q}$, we make a brief sortie into the discovery-or-invention controversy. The question of whether we invent or discover mathematical entities is a happy hunting ground for philosophers which is occasionally invaded by mathematicians in need of a break from their proper business. A definitive answer seems just as far away now as when the question was first raised, and it wouldn't change the fact that mathematics is much more exciting than philosophy, anyway! In the context of this chapter, the issue is about whether the non-rational reals have been sitting around 'in nature' ever since the world began, waiting to be discovered, or we have invented them in order to make highly artificial problems such as 'find a number whose product with itself is 5' have (equally artificial) solutions. Most answers tend to be along the lines, 'In a certain sense we invent (or discover) mathematics...', and it is in explaining exactly *what* sense that philosophical controversy can thrive. But, as far as irrational numbers are concerned, we believe that there are strong grounds for taking an invention stance.

First, if we think of numbers in the 'how many'? sense and consider the sorts of things we do with this idea, it is fairly clear that rational numbers suffice for our daily life. If we try to argue that irrational numbers crop up when we want to measure continuous things such as lengths and volumes, it could be pointed out that continuity (and, hence, the need for $\sqrt{5}$) is an invented concept, anyway: representations of real lines are not continuous, for they consist of discrete collections of ink molecules—even less continuous than the rationals! So, if we claim to be capturing real live irrational numbers in our rationally constructed nests, it must be pointed out that the argument concerns not the nature of the line but the nature of our mental image of it. What we do know for certain is that $\sqrt{5}$, whatever that symbol means, *cannot* mean a rational number. The discovery lobby could then argue that we have discovered points on the line which are not rational. The opposition could reply that our so-called discovery only amounts to the following: that by associating equivalence classes of nests with a new type of number we can build a new system (whose elements we choose to call numbers) having nothing ostensibly to do with geometry but which nevertheless has a property which is very close to our intuitive idea of continuity. But, of course, this association is tantamount to invention!

A second point, and one which really clinches it for invention, is the very unreliability of our number line intuition. The sort of intuition we used to get a point in all the intervals of a nest didn't rely upon the intervals being closed, so we could have carried out the same thought-experiment with open intervals. We may be encouraged to believe that this won't make any difference by the fact that just leaving out the two end-points still leaves infinitely many points between. *But it makes an enormous difference*: the sequence $(0, 1), (\frac{1}{3}, \frac{2}{3}), (\frac{2}{5}, \frac{3}{5}), (\frac{3}{7}, \frac{4}{7}), \ldots$ clearly has $\frac{1}{2}$ as a common member, but the translated sequence $(0, 1), (0, \frac{1}{3}), (0, \frac{1}{5}), (0, \frac{1}{7}), \ldots$ has an empty intersection. What incredible luck that we chose to work with nests of *closed* intervals!

The real message from all this is that we have invented a system (or we will have done so by the end of the book) which, although motivated by a geometric intuition about points and lines, is far richer than the geometric line. We aim to scatter throughout this book properties of $\mathbb{R}$ which can only be described as mind-boggling, and no amount of careful thinking about lines would have revealed them. Even hindsight is not enough. You may decide, for example, that the trouble with open intervals is that they are just that little bit smaller than the corresponding closed intervals, and since we are talking about sequences of intervals whose lengths are approaching zero, that little bit may be enough to tip the balance. We can knock that idea on the head in a most spectacular manner: suppose that we work with intervals which are infinite in extent—for example, $[0, \infty)$, $[1, \infty)$, $[2, \infty)$, $[3, \infty), \ldots$. (Don't worry about the mystique of '$\infty$'—the notation $[a, \infty)$ just means the set of all points $\geq a$.) These intervals are certainly nested, but this intersection is also empty, and this has nothing to do with the open/closed issue or with 'getting small too fast', because the intersection is empty irrespective of whether the left-hand end-points are open or closed, and the intersection is empty even though the intersection of all the intervals *up to any particular interval in the sequence* is an infinite interval!

All the fun of analysis can be traced to its dependence on *infinite* processes. We think we know about the line, so it is a bit of a shock to be mind-boggled by it. But at least we suspect before we start that infinity is likely to be slippery, so, to make sure our fun is safe and disciplined, we aim to grasp infinity with finite tools (which are all we have) and elect someone such as Logical Validity to be umpire of all our games. Before these metaphors collapse, let us return to the real numbers and mention two other ways of capturing them.

### 4.6.3 Route 2: Monotonic Sequences

Consider the sequence of rationals, $0, \frac{1}{2}, \frac{2}{3}, \frac{3}{4}, \frac{4}{5}, \ldots$ whose $n$th term is $(n-1)/n$. Its significant properties for our purposes are (a) each term is bigger than its predecessor; and (b) although the terms get progressively bigger, they remain within finite bounds—for example, no term of the sequence is bigger than 3. A third property is also clear: (c) the sequence is heading towards an ultimate goal of 1 in the sense that, whichever positive rational number $\varepsilon$ you care to specify, however small, I can always find some term of the sequence beyond which all the terms differ from 1 by less than $\varepsilon$. In this example it is very easy, for if you specify, say, $10^{-6}$, it is clear that all terms after the millionth are within $10^{-6}$ of 1.

This idea can be phrased in terms of the number line. We have a sequence of points for which (a) each point is to the right of its predecessor and (b) there is some point on the line with no points of the sequence to the right of it. Property (c) says that there is an 'ultimate point' P with the property that, if you specify any distance $\varepsilon$ (other than zero), then all the sequence points after the $N$th (where $N$ generally depends on $\varepsilon$) lie within $\varepsilon$ of P.

Sequences which satisfy (a) (with 'bigger than' relaxed to 'bigger than or equal to') are called monotonic increasing sequences and their 'ultimate goals' are called limits. Property (b) is usually expressed by saying the sequence is bounded above. The vital intuition which this approach seizes upon is that any sequence of points satisfying (a) and (b) must also satisfy (c). Transferring this belief about points to numbers, the aim is to define each real number $r$ as the set of monotonic increasing sequences of rationals which have $r$ as their limit. Well, that's the idea, but, of course, it won't do as it stands, because we have, as yet, no way of specifying $r$ except by reference to some sequence of rationals, so we need some method of saying that two sequences have the same limit, but only mentioning *rational* numbers. One way of doing this is to look at the difference between the sequences. Let $a_1, a_2, a_3, \ldots$ and $b_1, b_2, b_3, \ldots$ be two monotonic increasing bounded sequences of rationals, and for each $i$ let $c_i = a_i - b_i$. Then saying that the first two sequences have the same limit is equivalent to saying that $c_1, c_2, c_3, \ldots$ has zero as its limit, and 0 and all the $c_i$'s are rational! Of course, the c-sequence won't necessarily be monotonic, but our definition of a limit still makes sense.

This approach is very similar to route 1, because the sequences of left-hand end-points of the nest intervals are precisely bounded monotonic increasing sequences of rationals.

### 4.6.4 Route 3: Dedekind Cuts

This approach, published by the German mathematician Richard Dedekind in 1872, is rather different from the other two we have mentioned. It very nearly captures that feature of the line which $\mathbb{Q}$ doesn't have—namely, continuity. We first need to remark that sets of numbers, even if bounded, need not have least or greatest members. For example, the set of rational numbers whose squares do not exceed 5 is certainly bounded above, but it has no greatest member. [**4.13**] Even simpler is the set of numbers bigger than 1. This has no greatest member nor has it a least member, for different reasons.[**4.14**]

Now the set of rational points $\mathbb{Q}$ has gaps in the following sense: suppose that we split $\mathbb{Q}$ into two parts $\mathscr{L}$ and $\mathscr{R}$, as in Figure 4.8, so that $\mathscr{L}$ neither is empty nor includes all of $\mathbb{Q}$, it has no greatest member, and it has the property that given any rational $l$ in $\mathscr{L}$ all the rationals less than $l$ are also in $\mathscr{L}$. $\mathscr{R}$ is just all the *rest* of $\mathbb{Q}$ (or the *right*-hand bit). An example of an $\mathscr{L}$ satisfying all these conditions is $\mathscr{L} = \{x : x \in \mathbb{Q}, x < 0 \vee x^2 < 5\}$. In this case $\mathscr{L}$ has no greatest member and $\mathscr{R}$ has no least member [**4.15**]—a situation which it is fair to describe graphically as a gap in $\mathbb{Q}$. Now let us imagine what would happen if we used the same idea to split not $\mathbb{Q}$ but the whole line into an $\mathscr{L}$ and an $\mathscr{R}$. We think of $\mathscr{L}$ as the set of all points up to but not including a specific point P, and $\mathscr{R}$ is the rest of the line. Then $\mathscr{L}$ has no right-most member but $\mathscr{R}$ has a left-most member, P. In other words, the *continuity* of the line rules out the possibility that $\mathscr{L}$ could have no right-most member and $\mathscr{R}$ no left-most member.

The idea of Dedekind's construction is to think of the set of real numbers as the set of all $\mathscr{L}$-type subsets of $\mathbb{Q}$. Those for which the corresponding $\mathscr{R}$

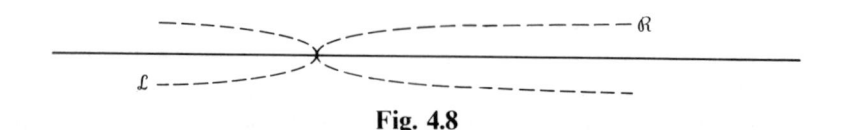

**Fig. 4.8**

has a least element are associated with rational numbers; the others, with irrationals.

All these methods of generating $\mathbb{R}$ from $\mathbb{Q}$ have sought to invent a system which is free of properties perceived as shortcomings of $\mathbb{Q}$. Nests with rational end-points don't necessarily home in on rational points; nests with real end-points do home in on real points. Bounded monotonic sequences of rationals don't always have rational limits; bounded monotonic sequences of reals do have real limits. The rationals have gaps; the reals don't. And so on.

We are still quite some way from proving these claims, but this chapter has been hard work (often hard thought rather than hard mathematics), and we are now at a 'natural break' point, anyway. The theme of the real numbers will be taken up again in Chapters 7, 8, 10 and 11, but enough has been said in this chapter to indicate, first, why it is desirable to have a purely arithmetic description of a geometric continuum, and second, that the translation of geometric ideas to arithmetic language sufficiently precise to be useful is surprisingly difficult and subtle.

# 5 A Variety of Versions and Uses of Induction—in which another triviality plays the lead

'Mathematicians are like lovers... grant a mathematician the least principle, and he will draw from it a consequence, which you must also grant him, and from this consequence another.'

Fontenelle

## 5.1 INDUCTION TO INDUCTION

TWEEDLEDEE: I've started to educate myself, Alice, as you suggested. I found a little book in the Red Queen's library by some chap called Fibonacci. They had very quaint ways of describing themselves in those days: this book was... 'by Leonardo, the everlasting rabbit breeder of Pisa'.
ALICE: Certainly unorthodox. What attracted you to it?
DEE: Oh, I just thought the arithmetic would appeal to you. His rabbits all had very tidy habits. They were all young for one month, middle-aged for the next and old from then on. Each male/female pair stayed together all their lives and produced one new male/female pair at the start of every month of their old age.
WHITE RABBIT: Rubbish! We rabbits may be well organised but we don't behave like clockwork robots. It sounds as if your friend Fibonacci was really a mathematician and didn't want to admit it for some reason—unpopular with the government, probably. I'll bet the only rabbit he ever thought about was *Oryctolagus hypothetico*!
ALICE: Yes, all right, but that may not be a criminal offence if the arithmetic is interesting. Where *does* the arithmetic come from, Dee?
DEE: Well, Fibonacci started with just one pair of new-born rabbits, and he wanted to know how his population would grow month by month. I found this table showing the number of pairs at the start of each month:

| Start of month number | 1 | 2 | 3 | 4 | 5 | 6 | 7 | 8 |
|---|---|---|---|---|---|---|---|---|
| Number of pairs | 1 | 1 | 2 | 3 | 5 | 8 | 13 | 21 |

TWEEDLEDUM: Most interesting. There seems to be a pattern; isn't the pair-population in any month just the sum of the populations in the two previous months? $f_n = f_{n-1} + f_{n-2}$ if you want to be formal and call $f_n$ the number of pairs at the start of month $n$.

ALICE: Does the pattern carry on like that? Are there 34 pairs at the start of the ninth month?
DEE: I don't know. He didn't give any more terms of the sequence.
ALICE: Well, how did he work it out? We could use the same method to carry on.
DEE: He did explain the method but the Red Queen has removed those pages—something to do with her censorship principle of too much knowledge too easily obtained rots the brain.
DUM: Right, find Fibonacci and ask him. You said he was everlasting and this *is* Wonderland, so he should still be around.
DEE: No, I didn't say that, but I can see why you thought I did. I'll have to write it down to explain to you. You thought I said

LEONARDO, THE EVERLASTING (RABBIT BREEDER)

but what I meant was

LEONARDO, THE (EVERLASTING RABBIT) BREEDER

DUM: Hmm, useful things, brackets. Pity we don't say them as well as write them. You've answered something else that was bothering me: we would have needed to know how long the rabbits lived.

[The friends now had two problems: how were the numbers in the sequence worked out? [**5.1**] and why is $f_n = f_{n-1} + f_{n-2}$ the right connection between any three consecutive terms? [**5.2**] We join them again after they have solved these and have become interested in the sequence itself rather than the rabbits.]

DEE: I found this very addictive. It kept me awake all night, and I'm seriously considering suggesting to the Red Queen that her library display a government health warning. They really did breed like rabbits! I just couldn't stop calculating—here's what I've got so far:

| 1 | 1 | 2 | 3 | 5 | 8 | 13 | 21 |
|---|---|---|---|---|---|---|---|
| 34 | 55 | 89 | 144 | 233 | 377 | 610 | 987 |
| 1597 | 2584 | 4181 | 6765 | 10946 | 17711 | 28657 | 46363 |
| 75025 | 121393 | 196418 | 317811 | 514229 | 832040 | | |

You'll notice I've underlined all the terms which are divisible by 5.
DUM: Why did you do that?
DEE: Just tell me which terms they are, then you may see.
DUM: The 5th, 10th, 15th, 20th, 25th and 30th.
DEE: Exactly! Every fifth term is a multiple of 5.
ALICE: You mean every one so far.
DEE: Yes, but you must agree it looks pretty suspicious.
ALICE: Agreed, but I've been caught out before. I'd like to be sure. Work out some more.

[They work out the next five but it takes a long time. The numbers are getting bigger, the sums harder; the trio are getting tired, they are making mistakes and have to check everything very carefully. At last they get there, and, sure enough, the 35th term is also divisible by 5.]

ALICE: I'm a bit happier about it now, but not convinced.
DUM: No, we don't seem to have any guarantee that $f_{40}$ is going to be a multiple of 5, and it will be even harder to check that.
ALICE: You're right, but it's even worse than that: if I want proof that $f_{5n}$ is always divisible by 5 instead of just a strong feeling in my bones that it's true, it won't matter whether you check the first five hundred terms or the first five million. And if you do check the first five million for me and it works, my bones won't be able to feel the difference when you do the next five terms. It isn't only economics that has a law of diminishing returns.

[There are not many engineers in Wonderland, but March Hare has dabbled in the field, and he comes up with the next breakthrough.]

MARCH HARE: If I want to predict how a machine will work, what do I need to know? What it's made of and how the bits are put together, that's all. Fibonacci's sequence is like that. The bits are the two 1s at the start, $f_1$ and $f_2$, and the whole thing is put together by adding: $f_1 + f_2$ gives you $f_3$, $f_4$ comes from $f_2 + f_3$, and so on. If we can't decide from that whether Alice's bones are reliable, then we can't do it at all!
ALICE: You mean that all this tedious arithmetic we've been doing isn't really necessary.
MARCH HARE: That's right. Let's say you've already worked out $f_{30}$, 832 040 wasn't it?—certainly divisible by 5. Now, what do you want to know about $f_{35}$?
ALICE: What it is, of course, so that I can check by dividing it by 5.
MARCH HARE: No, you don't. You need far less information than that. Suppose I told you I was thinking of a number which was 50 more than some multiple of 5. You certainly wouldn't be able to say what the number was, but you could be quite certain that it was a multiple of 5.
ALICE: Yes, I see your point. How does it help?
MARCH HARE: Let's take $f_{30}$, 832 040. We hope $f_{35}$ is a multiple of 5, too, but we want to prove it without doing any more arithmetic. We could work out $f_{31}$ from $f_{29}$ and $f_{30}$, but let's not bother. Just call it $x$ and see what happens:

$f_{30} = 832\,040$
$f_{31} = x$
$f_{32} = x + 832\,040$
$f_{33} = (x + 832\,040) + x = 2x + 832\,040$
$f_{34} = (2x + 832\,040) + (x + 832\,040) = 3x + 2 \times 832\,040$

$$f_{35} = (3x + 2 \times 832\,040) + (2x + 832\,040)$$
$$= 5x + 3 \times 832\,040$$

Now you can see that we don't need to know what $x$ is. $f_{35}$ is obviously a multiple of 5.

DUM: Very neat, March Hare, but this still only tells us that $f_{35}$ is a multiple of 5. What about $f_{40}, f_{45}, \ldots$ and all the rest?

MARCH HARE: Good question, Dum, and your objection is quite proper. My example was only an illustration of my general theorem, which modesty was preventing me from expounding.

DUM: Most commendable modesty, March Hare, but I think we really need your theorem. What does it say?

MARCH HARE: It refers to *any* sequence of integers which obeys the rule $f_n = f_{n-1} + f_{n-2}$, and says:

$$(\forall n)(f_n \text{ is a multiple of } 5 \Rightarrow f_{n+5} \text{ is a multiple of } 5)$$

DUM: Presumably you've called it a theorem because you can prove it?

MARCH HARE: Yes, but the proof is quite simple, nothing to be proud of. Having the idea in the first place was the hardest bit. Still, let's do things properly. Here's my proof. It uses the direct method (Chapter 1, in case you've forgotten).

Call $f_n$ $5t$, where $t$ is some integer, just to remind us that it is divisible by 5, and call the next term $x$. We can carry on now:

$$5t, \ x, \ x + 5t, \ 2x + 5t, \ 3x + 10t, \ 5x + 15t$$

and $5x + 15t$ is the term $f_{n+5}$, and it is clear that it's a multiple of 5.

Now go back to Fibonacci's sequence, We've worked out that $f_5 = 5$, which is certainly a multiple of 5; my theorem then *guarantees* that $f_{10}$ is also a multiple of 5 without any further arithmetic. Apply my theorem again to see that the same is true of $f_{15}$, and then $f_{20}$, and $f_{25} \ldots$, and so on, as far as you like.

At this point March Hare modestly receives rapturous applause and the friends adjourn to a tea party to celebrate the Red Queen's failure to keep them in ignorance of mathematics. We'll leave them there and take a sober look at what has been achieved.

Let us denote by $S_k$ the statement '$f_{5k}$ is divisible by 5'. Then what Alice felt in her bones was that all the statements $S_1, S_2, S_3, \ldots$ were true. The way that March Hare provided her with the proof consisted in showing two distinct things, which we shall call $I_1$ and $I_2$. $I_1$ is '$S_1$ is true' and $I_2$ is '$(\forall k)(S_k \Rightarrow S_{k+1})$', and this is called the method of mathematical induction. $I_2$ is sometimes called the *inductive step* and $S_k$ the *induction hypothesis*. (We attempt to deduce the truth of $S_{k+1}$ on the hypothesis that $S_k$ is true.) $I_1$ is called the *basis* of the induction. (Once we know that $S_1$ is true, we can be sure of $S_2$, because $S_1 \Rightarrow S_2$, of $S_3$ because $S_2 \Rightarrow S_3, \ldots$, and so on.)

Here are four further examples of this basic principle. We have deliberately written out the first one in full detail, and in each subsequent example have decreased the amount of detail until Example 4 approaches the rather terse style encountered in many textbooks.

**Example 1**  No pair of consecutive terms in the Fibonacci sequence have a common factor (other than 1, of course).

To make this amenable to induction, we first have to decide what the sequence of statements $S_1, S_2, S_3, \ldots$ is. If the Fibonacci sequence is $f_1, f_2, f_3, \ldots$, then $S_1$ is '$f_1$ and $f_2$ are coprime' (coprime means having no common prime factor), $S_2$ is '$f_2$ and $f_3$ are coprime',... and in general $S_k$ is '$f_k$ and $f_{k+1}$ are coprime'.

Now $S_1$ is clearly true, since $f_1$ and $f_2$ are both 1, so this deals with the basis for the induction.

Can we now, for arbitrary $k$, show that $S_{k+1}$ follows from $S_k$? Well, $S_k$ says that $f_k$ and $f_{k+1}$ are coprime, and we want to show that $f_{k+1}$ and $f_{k+2}$ are coprime. $f_{k+2} = f_k + f_{k+1}$ from the definition of the Fibonacci sequence, so if $f_{k+1}$ and $f_{k+2}$ had a common prime factor, this would also be a common factor of $f_k$ and $f_{k+1}$. (Prove this. [5.3]) $S_k$ says that no such prime factor exists.

So we have shown $S_k \Rightarrow S_{k+1}$ by establishing the contrapositive (see Chapter 1) $\sim S_{k+1} \Rightarrow \sim S_k$. We have done this for a perfectly arbitrary $k$, so we have: $(\forall k)(S_k \Rightarrow S_{k+1})$.

So $I_1$ and $I_2$ are both done, and by induction all the $S_k$ are true. That is, every pair of consecutive Fibonacci numbers are coprime.

**Example 2**  The sum of the first $n$ natural numbers is $\frac{1}{2}n(n+1)$.

*Proof* Let $S_k$ be the statement '$1+2+3+\cdots+k = \frac{1}{2}k(k+1)$.' $S_1$ is clearly true, because it says $1 = \frac{1}{2}(1)(1+1)$. $S_{k+1}$ says

$$'1+2+3+\cdots+k+(k+1) = \frac{1}{2}(k+1)(k+2)'$$

Now
$$1+2+3+\cdots+k+(k+1) = \frac{1}{2}k(k+1)+(k+1) \quad (1)$$
on the hypothesis that $S_k$ is true

We want to show that this is $\frac{1}{2}(k+1)(k+2)$ to establish $S_{k+1}$. The bit of algebraic manipulation we need is virtually obvious because the expression we have for $1+2+3+\cdots+(k+1)$ in line (1) above, and the one we want, both contain $k+1$ as a factor. So, extracting this common factor and continuing from line (1):

$$\vdots$$
$$= (k+1)(\tfrac{1}{2}k+1) = \tfrac{1}{2}(k+1)(k+2)$$

as required.

We have shown that each $S_{k+1}$ follows from its predecessor $S_k$, so all the $S_k$ are true by induction.

**Example 3** For each natural number $n$, the derivative of $x^n$ is $nx^{n-1}$.
*Proof* Let $S_k$ be 'the derivative of $x^k$ is $kx^{k-1}$.' $S_1$ is clearly true. We now need to show that $S_k \Rightarrow S_{k+1}$—that is, that if

$$\frac{d}{dx}(x^k) = kx^{k-1}$$

it follows that

$$\frac{d}{dx}(x^{k+1}) = (k+1)x^k$$

We write $x^{k+1}$ as $x \times x^k$ and use $S_1$ and the product rule for differentiation:

$$\frac{d}{dx}(x^{k+1}) = \frac{d}{dx}(x \times x^k) = \left[x \times \frac{d}{dx}(x^k)\right] + \left[\frac{dx}{dx} \times x^k\right]$$
$$= (x \times kx^{k-1}) + (1 \times x^k)$$
$$= kx^k + 1x^k = (k+1)x^k$$

Hence, the result is true by induction.

**Example 4** An $n$-element set has $2^n$ subsets (counting the empty set $\emptyset$, and the set itself).
*Proof* The only subsets of $\{x\}$ are $\{x\}$ and $\emptyset$, and so the result is true for $n=1$.

Now let $A$ be a set of $k+1$ elements, where $k$ is any natural number, and let $a$ be some particular member of $A$.

Each subset of $A$ either contains or does not contain $a$, so each subset is of the form $S \cup \{a\}$ or $S$, where $S$ is a subset of $A \setminus \{a\}$.

Assuming that $k$-element sets have $2^k$ subsets, there are $2^k$ subsets of $A$ of the first type, and the same number of the second—a total of $2 \times 2^k = 2^{k+1}$ subsets of $A$.

Here are some similar examples to try.

[5.4] Prove that
$$(1 \times 5) + (2 \times 6) + (3 \times 7) + \cdots + (n \times (n+4)) = \tfrac{1}{6}n(n+1)(2n+13)$$

[5.5] Prove that
$$(1 + \tfrac{1}{2})(1 + \tfrac{1}{4})(1 + \tfrac{1}{16}) \cdots \left(1 + \frac{1}{2^{(2^{n-1})}}\right) = 2\left(1 - \frac{1}{2^{(2^n)}}\right)$$

[5.6] Work out a few values of $4^n - 3n - 1$ ($n \in \mathbb{N}$). Guess a number which is a factor of all the values of $4^n - 3n - 1$ and prove by induction that your guess is correct.

[5.7] Show by induction that, provided that $x \neq 0$,
$$\frac{d}{dx}(x^{-n}) = -nx^{-n-1}$$

for all $n \in \mathbb{N}$, assuming that
$$\frac{d}{dx}\left(\frac{1}{x}\right) = -\frac{1}{x^2}$$

## 5.2 VARIATIONS ON THE THEME

### 5.2.1 Induction Need not Start at $n = 1$

Prove that the sum of the interior angles of a convex polygon with $n$ sides is $(n - 2) \times 180$ degrees.

To formulate this as an induction problem, it is tempting to let $S_k$ be the statement 'Convex $k$-gons have an internal angle sum of $(k - 2) \times 180°$.' That's fine, provided that we notice that $k = 3$ is the smallest value for which $S_k$ makes geometric sense (and $S_3$ happens to be true!) The demonstration that $S_k \Rightarrow S_{k+1}$ for all $k \geq 3$ is fairly easy for convex polygons because if we cut off a corner, as in Figure 5.1, the convex $(k+1)$-gon $A_1 A_2 A_3 \ldots A_{k+1}$ is split into a convex $k$-gon $A_1 A_3 A_4 A_5 \ldots A_{k+1}$ and the triangle $A_1 A_2 A_3$.

We need to show that
$$\theta_1 + \theta_2 + \theta_3 + \theta_4 + \cdots + \theta_{k+1} = (k-1) \times 180°$$

# A VARIETY OF VERSIONS AND USES OF INDUCTION

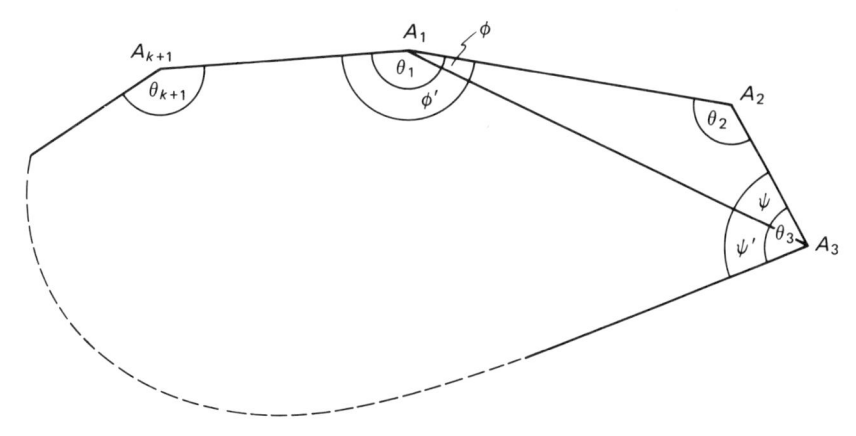

Fig. 5.1

But the left-hand side of this equation is

$$= (\phi' + \psi' + \theta_4 + \theta_5 + \cdots + \theta_{k+1}) + (\phi + \theta_2 + \psi)$$
$$= (k-2) \times 180° \quad + \quad 180°$$

(by the induction hypothesis $S_k$, and $S_3$)

$$= (k-1)180°$$

as required.

A summary of all this is that we have shown:

$$I'_1: \text{`}S_3 \text{ is true'}$$

and

$$I'_2: \text{`}S_k \Rightarrow S_{k+1} \text{ for all } k \geq 3\text{'}$$

and the obvious inductive conclusion from this is:

$$S_k \text{ is true for all } k \geq 3$$

Sometimes a statement about a general natural number $n$ (a predicate over $\mathbb{N}$) may make perfectly good sense for all $n$, but just happens to be false for the first few values of $n$ and true for the rest. For example: find the values of $n$ for which $n! > n^3$ and prove your result by induction. [**5.8**]

Finally, investigate when $2^n - 1$ is divisible by 3, make a general conjecture and prove it inductively. [**5.9**]

## 5.2.2 A Different Way of Looking at Induction

$\mathbb{N}$ has the following very obvious property: any non-empty subset of $\mathbb{N}$ has a least element. This may be obvious, but it does distinguish $\mathbb{N}$ from many other mathematically significant sets, for the same property does not hold if we substitute $\mathbb{R}, \mathbb{Z}, \mathbb{Q}$ or even the positive reals or rationals for $\mathbb{N}$. So what? Well, that extremely trivial observation about $\mathbb{N}$ provides us with an alternative justification for induction (in case you happen to be sceptical about its validity), and with an alternative version of the inductive principle which is often the one used in practice.

First, the justification. Suppose that we know that $S_1$ is true, and have shown $(\forall k)(S_k \Rightarrow S_{k+1})$ but are still uneasy about the use of the 'and so on' bit to infer $(\forall n)(S_n)$. Let us suppose $\sim(\forall n)(S_n)$ and aim to derive a contradiction. Our supposition that $S_n$ is not always true means that there are some (at least one) $n$ for which $S_n$ is false. Let $F$ be the set of all $n$ for which $S_n$ is false, so that $F$ is a non-empty subset of $\mathbb{N}$. It therefore has a *least* member $l$ which is greater than 1. So $S_l$ is false and $S_{l-1}$ is true.

But $S_{l-1} \Rightarrow S_l$, so $S_l$ is true and we have our contradiction. Hence, $(\forall n)(S_n)$.

[5.10] Why is $l > 1$ and why do we require this?

What the above argument really amounts to is a demonstration that the validity of proof by induction follows from the validity of proof by contradiction and the least element principle for $\mathbb{N}$.

For our first illustration of this version of induction in action, we give a proof that $\sqrt{5}$ is irrational by a method quite different from those given in Chapter 4 and Chapter 7.

Assume that $\sqrt{5}$ is rational. It is positive, so can be written as $a/b$, where $a$ and $b$ are natural numbers. Now there is no uniqueness about $a$ or $b$, so consider the set $S$ of all natural numbers $b$ which are denominators of fractions representing $\sqrt{5}$. $S$ is then a non-empty subset of $\mathbb{N}$, so has a least member $l$. Hence, $\sqrt{5} = n/l$, where $n$ is some other natural number.

We pause here to anticipate a rather nice feature of this proof. If we assumed the Fundamental Theorem of Arithmetic (Chapter 2), we could obtain $l$ simply by taking any naturals $x, y$ for which $\sqrt{5} = x/y$, factorising $x$ and $y$ and cancelling all their common factors. However, our method will be independent of the fundamental theorem. If $\sqrt{5}$ *has* a rational expression, there must be a minimal $l$ and it is only the *existence* of $l$ we require, not any method of *finding* it.

Returning to our proof, we have $\sqrt{5} = n/l$, so $\sqrt{5}(l)$ is a natural number and $l$ is the least natural number for which this is true.

Now $(3 - \sqrt{5})l$ is also a natural number, $k$, which is less than $l$. Prove this. [5.11]

But $k\sqrt{5}$ is also a natural number (prove this [5.12]), which contradicts the minimality of $l$, and the proof is done.

[5.13] Apply the same method to show that $\sqrt{94}$ is irrational.

[5.14] If you tried to use the same method to show that $\sqrt{49}$ was irrational, where would it break down?

[5.15] Generalise these examples to show that any natural number which does not have a natural number for its square root must have an irrational square root.

As mentioned in Chapter 2, our proof there of the Fundamental Theorem of Arithmetic worked by this disguised form of induction. We postulated a least natural number not uniquely prime factorisable, then produced a smaller one!

### 5.2.3 Weak and Strong Induction

Consider the Lucas sequence; 1, 3, 4, 7, 11, 18, 29, 47, ..., which is constructed in the same way as Fibonacci's—namely $l_n = l_{n-1} + l_{n-2}$ for all $n \geq 3$, but the first two terms are 1 and 3 instead of 1 and 1.

The property we would like to prove for the Lucas sequence is $(\forall n)(l_n \leq (\frac{7}{4})^n)$.

If $S_k$ is '$l_k < (\frac{7}{4})^k$' it is clear that $S_1$ is true. To complete the induction, can we show that $S_k \Rightarrow S_{k+1}$?

$$l_{k+1} = l_k + l_{k-1} < (\tfrac{7}{4})^k + l_{k-1} \tag{1}$$

and we want to show that this is less than $(\frac{7}{4})^{k+1}$. In order to do this, we clearly need some information about the size of the $l_{k-1}$, and our induction hypothesis $S_k$ does not provide this.

Let us speculate a little and assume for the moment that both $S_k$ and $S_{k-1}$ are true; in other words, $l_k < (\frac{7}{4})^k$ and $l_{k-1} < (\frac{7}{4})^{k-1}$. Returning to line (1) above, this would allow us to deduce

$$l_{k+1} < (\tfrac{7}{4})^k + (\tfrac{7}{4})^{k-1} = (\tfrac{7}{4})^{k+1}(\tfrac{4}{7} + \tfrac{16}{49}) = (\tfrac{7}{4})^{k+1}(\tfrac{44}{49}) < (\tfrac{7}{4})^{k+1}$$

That is, $S_{k+1}$ is true.

You can check that $S_1$ and $S_2$ are both true; then we can write down formally what we have shown so far:

$I''_1$: '$S_1 \wedge S_2$'

$I''_2$: '$(\forall k > 1)((S_{k-1} \wedge S_k) \Rightarrow S_{k+1})$'

From these it is fairly clear that we can deduce $(\forall n)(S_n)$, as required. Write out a formal argument to show this, along the lines of the argument used in subsection 5.2.2. [**5.16**]

In ordinary induction we only needed $S_k$ in order to deduce $S_{k+1}$, but in this version we need a *stronger* hypothesis, $S_{k-1} \wedge S_k$, in order to deduce the same, so we could describe the whole statement $S_{k-1} \wedge S_k \Rightarrow S_{k+1}$ as *weaker* than the inductive step in the original version.

Our next example takes this strategy of strengthening the induction hypothesis and correspondingly weakening the induction step to its extreme, while still retaining the basic idea of induction.

We prove that each natural number $n$ is the sum of distinct powers of 2. The result is clearly true for $n = 1$, since 1 is $2^0$. Now let $S_k$ be the statement that $k$ is a sum of distinct powers of 2 and let the induction hypothesis be $S_1 \wedge S_2 \wedge S_3 \wedge \cdots \wedge S_k$, from which we shall try to deduce $S_{k+1}$.

If $k + 1$ happens to be a power of 2, we are done. If not, let $2^m$ be the largest power of 2 which is less than $k+1$, so $k+1 = 2^m + w$, where $w < 2^m < k+1$. Why is $w < 2^m$? **[5.17]**

So $S_w$ is part of our induction hypothesis. Hence, $w$ is a sum of distinct powers of 2—all less than the $m$th power.

Therefore $2^m + w$ (i.e. $k+1$) is a sum of distinct powers of 2, and $S_{k+1}$ is proved.

This time we have shown

$$I'''_1: \ 'S_1'$$

$$I'''_2: \ '(\forall k)((S_1 \wedge S_2 \wedge S_3 \wedge \cdots \wedge S_k) \Rightarrow S_{k+1})'$$

and again we can deduce $(\forall n)(S_n)$ from this.

The importance of these two examples has been to show that induction is not quite a mechanical process, because great care must be taken to choose the induction hypothesis just strong enough to enable $S_{k+1}$ to be deduced.

Our final example to reinforce this message is the apparently simple task of proving that, for all $n \in \mathbb{N}$,

$$a_n = \frac{1 \times 3 \times 5 \times 7 \times \cdots \times (2n-1)}{2 \times 4 \times 6 \times 8 \times \cdots \times 2n} < \frac{1}{\sqrt{(3n)}}$$

the natural induction hypothesis is $S_k$: '$a_k < 1/\sqrt{(3k)}$'. $S_1$ is clearly true ($\frac{1}{2} < 1/\sqrt{(3 \times 1)}$), so let us try proving that $S_k \Rightarrow S_{k+1}$:

$$a_{k+1} = \frac{1 \times 3 \times 5 \times 7 \times \cdots \times (2k+1)}{2 \times 4 \times 6 \times 8 \times \cdots \times (2k+2)} = a_k \times \frac{2k+1}{2k+2} < \frac{1}{\sqrt{(3k)}} \times \frac{2k+1}{2k+2}$$

by the induction hypothesis.

We write the final expression on the right-hand side as

$$\frac{1}{\sqrt{[3(k+1)]}} \times \underline{\frac{\sqrt{[3(k+1)]}}{\sqrt{(3k)}}} \times \frac{2k+1}{2k+2}$$

and since we are aiming at proving this whole expression

$$< \frac{1}{\sqrt{[3(k+1)]}}$$

we hope that the underlined bit will turn out to be less than 1. You can check **[5.18]** that the underlined factor is

$$\sqrt{\left[\frac{4k^2 + 4k + 1}{4k^2 + 4k}\right]}$$

which is obviously not less than 1 and our attempt has failed.

Now we do something apparently perverse and try to prove an even more stringent condition—namely,

$$(\forall n)\left[a_n < \frac{1}{\sqrt{(3n+1)}}\right]$$

again with a natural induction hypothesis,

$$S_k: \;\! 'a_k < \frac{1}{\sqrt{(3k+1)}},$$

Now we try, as before, to prove $S_k \Rightarrow S_{k+1}$:

$$a_{k+1} = \frac{1 \times 3 \times 5 \times 7 \times \cdots \times (2k+1)}{2 \times 4 \times 6 \times 8 \times \cdots \times (2k+2)} = a_k \times \frac{2k+1}{2k+2} < \frac{1}{\sqrt{(3k+1)}} \times \frac{2k+1}{2k+2}$$

$$= \frac{1}{\sqrt{[3(k+1)+1]}} \times \sqrt{\underline{\left[\frac{3(k+1)+1}{3k+1}\right] \times \frac{2k+1}{2k+2}}}$$

Now manipulate the underlined factor and this time show that it really is less than 1 [**5.19**]—eureka!

How odd, that although

$$(\forall n)\left[a_n < \frac{1}{\sqrt{(3n)}}\right]$$

is a weaker assertion than

$$(\forall n)\left[a_n < \frac{1}{\sqrt{(3n+1)}}\right]$$

it is easier to prove the stronger one!

This phenomenon was brought to the attention of the mathematical public by George Polya in his book *Mathematics and Plausible Reasoning*, Volume 1, with the associated message that if a proof doesn't quite work by induction, it is sometimes worth trying to prove something stronger.

Here is a similar example. [**5.20**] Try to prove that whenever $x > 0$, $(\forall n)(1 + x)^n \geq nx)$ using the natural induction hypothesis $S_k: \;\! '(1 + x)^k \geq kx'$. Explain why it doesn't work, find a stronger $S_k$ and try again.

### 5.2.4 Backwards Induction

To illustrate this variation, we return to Fibonacci. We claim that every natural number can be written as a sum of *distinct* Fibonacci numbers. For example:

$17 = 8 + 5 + 3 + 1$
$100 = 89 + 8 + 3$
$50\,000 = 28\,657 + 10\,946 + 6765 + 2584 + 987 + 55 + 3 + 2 + 1$
$233 = 233$

We shall call such numbers fs numbers (Fibonacci summable).

What is clearly true is that our result holds for infinitely many numbers—namely the Fibonacci numbers themselves (for which the sum need only consist of a single term). The next step is to show that if $k$ is an fs number, then so is its predecessor $k - 1$. To do this, observe that the expression for $k$ as a Fibonacci sum is, or can be written as, a sum which either contains 1, or 2 but not 1. A couple of examples will make this obvious.

First, suppose that $k$ has been expressed as a sum in which the smallest term is 21. This 21 can be replaced by $8 + 13$; then the 8 by $3 + 5$; and finally, the 3 by $1 + 2$. The other case is illustrated when the smallest term is 34. We replace it by $13 + 21$; the 13 by $5 + 8$; and the 5 by $2 + 3$.

Now for $k - 1$. If we have written the sum for $k$ starting with 1, we just omit the 1 to get the sum for $k - 1$; and if the sum for $k$ starts with 2, we just replace the 2 by 1.

Here is a summary of what has been shown, using $S_k$ to stand for the statement '$k$ is an fs number'.

$I_1''''$: '$S_k$ holds for infinitely many $k$'

$I_2''''$: '$(\forall k > 1)(S_k \Rightarrow S_{k-1})$'

[5.21] Deduce $(\forall n)(S_n)$ from these.

[5.22] Try to find a proof of the result, using ordinary induction.

## 5.3 WHAT IS $n$?

This section contains a selection of examples in which the result to be proved is stated in terms of two or more natural numbers, or without mentioning any natural number explicitly, so that it is not clear how to choose $S_k$.

### 5.3.1 Fibonacci Again

The Fibonacci numbers satisfy the equation

$$f_{m+n} = f_{m-1}f_n + f_m f_{n+1} \text{ for all } n > 0 \text{ and all } m > 1$$

To attempt an inductive proof for this we seem to have two choices: take $S_k$ to be either

$$\text{`}(\forall n > 0)(f_{k+n} = f_{k-1}f_n + f_k f_{n+1})\text{'}$$

or

$$\text{`}(\forall m > 1)(f_{m+k} = f_{m-1}f_k + f_m f_{k+1})\text{'}$$

Let us see whether the former works. $S_2$ is

$$(\forall n > 0)(f_{2+n} = f_1 f_n + f_2 f_{n+1})$$

But $f_1$ and $f_2$ are 1, so this *is* true—it just reduces to a statement of the definition of the Fibonacci sequence.

Now can we derive $S_{k+1}$ from $S_k$? $S_{k+1}$ says

$$(\forall n > 0)(f_{k+1+n} = f_k f_n + f_{k+1} f_{n+1})$$

Starting with $f_{k+1+n}$, we argue:

$$f_{k+1+n} = f_{k+(n+1)} = f_{k-1}f_{n+1} + f_k f_{n+2}$$

(by $S_k$, replacing $n$ by $n+1$, which we are at liberty to do, because our induction hypothesis is that a certain relation involving $n$ and $m$ holds for a *particular m*, namely $k$, but for *arbitrary n*)

$$= (f_{k+1} - f_k)f_{n+1} + f_k(f_{n+1} + f_n)$$
$$= f_{k+1}f_{n+1} + f_k f_n$$

as required.

[**5.23**] In this case the alternative choice of induction hypothesis works too. Can you supply the details?

[**5.24**] There is a slicker proof, avoiding induction. Let $E_{m,n}$ denote the expression $f_{m-1}f_n + f_m f_{n+1}$; show that $E_{m,n} = E_{m-1,n+1}$, and apply this result repeatedly to obtain $E_{m,n} = E_{m-l,n+l}$. Take the case $l = m-2$ and this should give the required result.

## 5.3.2 The General Associative Law

To say that an operation—for example, multiplication—is associative is to claim that, for any three numbers $a, b$ and $c$,

$$a \times (b \times c) = (a \times b) \times c$$

In other words, when multiplying three numbers together, where we put the brackets makes no difference to the final answer. [There are plenty of operations which are not associative; for example, $(a \div b) \div c$ is generally very different from $a \div (b \div c)$.]

Because we are so familiar with multiplication, we generally assume that there is an automatic extension of associativity to cover the product of any number of numbers. That is, the product $a_1 a_2 a_3 \ldots a_r$ is unambiguous because it doesn't matter how we (mentally or physically) bracket it.

Now this happens to be true, but is it true just because the three-term associative property does imply the extension to $n$ terms, or is it true by virtue of additional properties peculiar to multiplication? The issue can be highlighted if you consider the operation $*$, defined by $a*b = a + b + ab$. You can easily check that $(a*b)*c = a*(b*c)$ [**5.25**], but it is by no means obvious that $(a*b)*((c*d)*e) = a*(b*(c*(d*e)))$ and checking it would be tedious.

So to our theorem:

If any operation $*$ is associative, then all bracketings of $a_1 * a_2 * a_3 * \cdots * a_n$ are equal.

The induction variable we take to be $n$, and we use the form of induction hypothesis discussed in Section 5.2.3.

$S_k$: All bracketings of strings of $k$ or fewer symbols are equal.

The basis of the induction is $S_3$, which holds because it just reiterates the fact that $*$ is associative.

(Expressions $a_1$ and $a_1 * a_2$ are unambiguous, anyway, so bracketing doesn't arise.)

Now consider a bracketed string of $k+1$ symbols. It is of the form:

(some bracketing of $(a_1*a_2*\cdots*a_p))*$(some bracketing of

$$(a_{p+1}*a_{p+2}*\cdots*a_{k+1}))$$

where $p$ is some integer between 1 and $k$ inclusive. Both brackets contain at most $k$ symbols, so by $S_k$ we can write the expression as

$$(\underline{a_1*(a_2*\cdots*a_p)})*(\underline{a_{p+1}*\cdots*a_{k+1}})$$

By $S_3$, applied to the three terms underlined, this can be rewritten as

$$a_1*((a_2*\cdots*a_p)*(a_{p+1}*\cdots*a_{k+1}))$$

Finally, apply $S_k$ again to the string following $a_1*$ to obtain

$$a_1*(a_2*(a_3*(a_4*\cdots*(a_k*a_{k+1})))\cdots)$$

That is, any bracketing of a string of length $k+1$ is equal to this standard bracketing, so all bracketings of strings of $k+1$ symbols are equal.

We have deduced $S_{k+1}$, so the induction is complete. □

### 5.3.3 Dots and Lines

Our next set of examples are concerned with graph (or network) theory, and they use induction through the least element principle discussed in Section 5.2.2.

For our purposes we deem a *graph* to be a finite non-empty set of points (the *vertices*). Some pairs of distinct vertices may be joined by lines, called the *edges* of the graph. The restriction to distinct vertices means that we specifically exclude loops (a vertex joined to itself by an edge), and we also exclude multiple edges (two vertices joined by more than one edge). Figure 5.2 should make the idea clear: (a) is a graph but (b) and (c) are not.

Finally, the graph (a) is said to have four components—the four separate bits labelled A, B, C and D (one of the components happens to have only one vertex).

The result we now prove is that, for any graph, $e+c \geq v$, where $e, c$ and $v$ are the numbers of edges, components and vertices, respectively.

The reason that the example lives in this section is that there is no '$n$'. Do we try induction on $e$, on $c$ or on $v$, or on some other parameter? We begin by showing that $e$ is a choice that works.

Let us suppose that $e+c \geq v$ does not hold for all graphs; let $G_0$ be such a graph, with the minimal possible number of edges $e_0$, (so $e+c \geq v$ for any graph with $e < e_0$), and let $c_0$ and $v_0$ be its numbers of components and vertices, respectively. Hence, $e_0 + c_0 < v_0$.

Now $e_0$ must be at least 1, for if not we would have $e_0 = 0$, $v_0 = c_0$ (each vertex would be a separate component), and then $e_0 + c_0 = c_0 = v_0$.

We can therefore delete one edge of $G_0$ to produce a new graph $G'$ having $v'$ vertices, $c'$ components and $e'$ edges. $v' = v_0$, $e' = e_0 - 1$, and $c' = c_0$ or

# A VARIETY OF VERSIONS AND USES OF INDUCTION

(a)   (b)   (c)

A

B

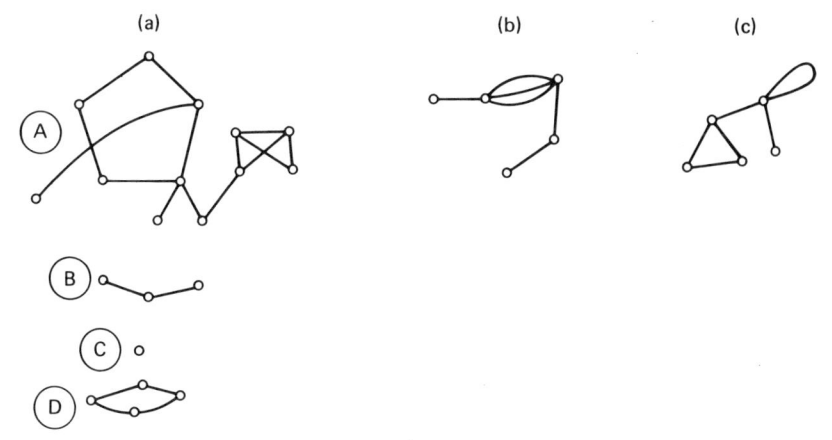

C

D

**Fig. 5.2**

$c_0 + 1$, since removal of an edge from a graph either leaves its component intact or splits it into two.

Hence, $e' + c' = e_0 - 1 + c_0$ or $e_0 + c_0$, so in either case

$$e' + c' \leqslant e_0 + c_0 < v_0 = v'$$

So $G'$ is also a graph for which $e + c \geqslant v$ is false, which is impossible, because $G'$ has fewer edges than $G_0$.

Hence, $e + c \geqslant v$ holds for all graphs. □

[**5.26**] Show that any connected graph (a graph with only one component) with $v$ vertices must have at least $v - 1$ edges. [*Hint*: Start with a totally disconnected graph—$v$ vertices, no edges—and build it up to a connected graph by adding edges. Consider what can happen to the number of components each time you add an edge.]

[**5.27**] Try induction on $c$ to prove $e + c \geqslant v$. (You may need the result of [**5.26**].)

[**5.28**] Try a proof by induction on $v$.

## 5.3.4 Examples from Logic

In Chapter 1 we saw how reasoning could be done in a symbolic language. If we restrict the language to those statements which do not require quantifiers, variables or predicates, we have a simpler system called the propositional calculus. This was the system discussed informally in the first part of Chapter 1. Its 'alphabet' consisted of the *connectives* $\Rightarrow$, $\Leftrightarrow$, $\vee$, $\wedge$ and $\sim$, *brackets* (and), and as many *letters* $a_1, a_2, \ldots$ as we required to stand for simple or *atomic* statements (those not requiring any connectives, such as 'Today is Tuesday' or 'Alice has two legs').

Compound statements (called *formulae* of the language) were strings of symbols taken from this alphabet, such as

$$(a_1) \vee ((a_2) \Rightarrow ((a_3) \wedge ((a_4) \Rightarrow (\sim (a_2)))))$$

but of course not every string is a formula. For example, $\Rightarrow (a_1 \vee \sim)(a_2) \sim))$ is meaningless, but if you try to give a watertight specification of which strings count as formulae, you may find it surprisingly difficult. Here induction comes to our aid, not to prove anything initially, but just to define 'formulae'.

Here is the definition:

(1) For each $i$, $a_i$ is a formula.
(2) If $F$ and $G$ are formulae, then so are $\sim(F)$, $(F) \vee (G)$, $(F) \wedge (G)$, $(F) \Rightarrow (G)$ and $(F) \Leftrightarrow (G)$.
(3) *Only* strings of the form specified by (1) and (2) above are formulae.

To see that this really does specify exactly which strings are formulae, imagine that you are confronted by a string. You first check that it is either a single letter or a string enclosed in brackets or preceded by $\sim$, or that it consists of two strings $F$ and $G$, each enclosed by brackets and joined by $\vee$, $\wedge$, $\Rightarrow$ or $\Leftrightarrow$. You then go through the same process with $F$ and $G$, and so on, to the bitter end. If at any stage the substring you are examining does not satisfy (1) or (2), then the whole string cannot be a formula. If they are satisfied right down to the level of single letters, then your string is a formula.

Our example above clearly fails the test at the first step, because no formula can begin with $\Rightarrow$.

A disadvantage of this very formal specification is that it forces formulae to have an excessive number of brackets. For example, $a_1 \wedge a_2$ doesn't qualify; it would have to be written as $(a_1) \wedge (a_2)$. One could add further rules to omit brackets when no ambiguity would arise, but although this would make formulae look simpler, it would complicate their definition.

In order to get to our example of an inductive proof, note that any formula is logically equivalent (defined in Chapter 1) to a formula whose only connectives are $\vee$, $\wedge$ and $\sim$. This is clear, because any occurrence of $(F) \Leftrightarrow (G)$ can be replaced by $((F) \Rightarrow (G)) \wedge ((G) \Rightarrow (F))$; then the $\Rightarrow$ can be removed by replacing $(F) \Rightarrow (G)$ by $(\sim (F)) \vee (G)$.

If $\mathscr{F}$ is a formula written without $\Rightarrow$ or $\Leftrightarrow$, and we transform it to a formula $\mathscr{F}'$ by replacing every $\vee$ by $\wedge$, every $\wedge$ by $\vee$, every $a_i$ by $\sim(a_i)$, then we claim that $\mathscr{F}'$ is logically equivalent to $\sim(\mathscr{F})$.

To prove it, we use induction on $n$, the *length* of $\mathscr{F}$, which is just the number of symbols in $\mathscr{F}$. Clearly, the result is true for $n = 1$, because $\mathscr{F}$ can then only be a single statement letter $a_i$. $\mathscr{F}'$ is $\sim(a_i)$, which is identical with $\sim(\mathscr{F})$.

Now assume its truth for all formulae of length $\leq k$ and let $\mathscr{F}$ be a formula of length $k + 1$. By our definition of a formula, $\mathscr{F}$ must be of the form $\sim(F)$, $(F) \vee (G)$ or $(F) \wedge (G)$, where $F$ and $G$ must have length $< k$.

In the first case
$$\mathscr{F}' = \sim(F') \equiv \sim(\sim(F)) = \sim(\mathscr{F})$$
using $\equiv$ for logical equivalence. In the second case
$$\mathscr{F}' = (F') \wedge (G') \equiv (\sim(F)) \wedge (\sim(G)) \equiv \sim((F) \vee (G)) = \sim(\mathscr{F})$$
The proof in the remaining case is similar, and the induction is complete.

[**5.29**] Check that we could just as easily have done an induction on the number of letter-occurrences in $\mathscr{F}$ (each letter contributing $l$ to the score if it occurs $l$ times).

[**5.30**] Show that if we tried an induction on the number of distinct letters in $\mathscr{F}$, then our method would fail.

[**5.31**] In the propositional calculus there are certain formulae which play a special role (like 0 and 1 in arithmetic). These are the tautologies, which are those formulae which come out as true in every line of their truth table. Some examples are $(P) \Rightarrow (P)$, $(\sim(P)) \vee (P)$ and $((P) \wedge ((Q) \wedge (\sim(P)))) \Rightarrow (R)$. Prove that a formula whose only connective is $\Leftrightarrow$ is a tautology if and only if each of its statement letters appears an even number of times. [*Hint*: You may find it helpful to show first that $\Leftrightarrow$ is an 'associative' and 'commutative' connective in the sense that $(P) \Leftrightarrow ((Q) \Leftrightarrow (R)) \equiv ((P) \Leftrightarrow (Q)) \Leftrightarrow (R)$ and $(P) \Leftrightarrow (Q) \equiv (Q) \Leftrightarrow (R)$.]

## 5.4 SOME CAUTIONARY TALES

In many of our examples induction was used to confirm a general result whose truth was suspected after examining some specific cases. Here are three situations which will lead you to plausible conjectures which happen to be false in general, and we invite you to suggest the conjectures.

[**5.32**] Examine the value of $x^2 + x + 41$ for various integer values of $x$. Is the result prime or composite?

[**5.33**] Place $n$ points on the circumference of a circle and join each point to every other point with a straight line. This will divide the circle into a number of regions. Place the $n$ points so that you achieve the maximal possible number of regions, $m$. Suggest a relationship between $n$ and $m$.

[**5.34**] For various natural numbers $n$, factorise $x^n - 1$ as far as possible into factors with integer coefficients. Make a conjecture about the size of these coefficients.

We conclude this lengthy chapter with three examples of abuse of induction, and invite you to spot what is wrong with each proof.

[**5.35**] For all $n$ the sum of the first $n$ natural numbers is $\frac{1}{2}(n + \frac{1}{2})^2$.
*Proof* Assume that the result holds for $n = k$. Then the sum of the first $k+1$ natural numbers is the sum of the first $k$, plus $k+1$. This is

$\frac{1}{2}(k+\frac{1}{2})^2 + k + 1$ and the proof will be complete if we can show that this is $\frac{1}{2}((k+1)+\frac{1}{2})^2$. This is easy, because

$$\frac{1}{2}(k+\tfrac{1}{2})^2 + k + 1 = \tfrac{1}{2}[(k+\tfrac{1}{2})^2 + 2k + 2]$$
$$= \tfrac{1}{2}[k^2 + 3k + \tfrac{9}{4}]$$
$$= \tfrac{1}{2}[(k+\tfrac{3}{2})^2]$$
$$= \tfrac{1}{2}[(k+1)+\tfrac{1}{2}]^2$$

[**5.36**] Next, a 'theorem' to please idle students but discourage you hard-working ones—but only if you believe it! For all $n$, if $n$ students take an examination, they must all achieve the same mark.
*Proof* The theorem holds trivially when $n = 1$.

Assume, for an induction hypothesis, that it holds for a group of students of size at most $k$, and consider a class $C$ of $k+1$ students.

Imagine two subclasses of $C$, $A$ and $B$, chosen so that one student, Tom, is in both $A$ and $B$, and everyone else in $C$ is put in $A$ or $B$ but not both.

Now let all of $C$ sit the same exam. $A$ and $B$ are necessarily of size at most $k$, so everyone in $A$ gets the same mark, which is Tom's mark. Similarly, everyone in $B$ gets the same mark and the induction is complete!

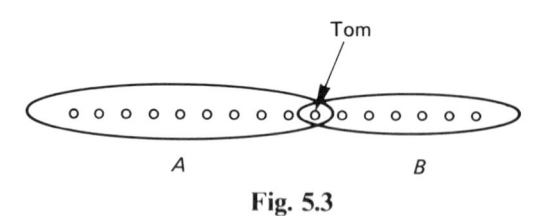

**Fig. 5.3**

For our final, rather subtle example, we need the idea of a graph (see Section 5.3.4) and a Hamiltonian circuit (a closed path in the graph which goes through every vertex once and once only). For example, in Figure 5.4, one of the graphs is Hamiltonian. Which one? [**5.37**]

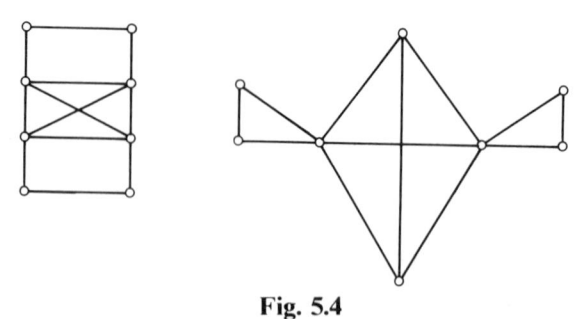

**Fig. 5.4**

**Dirac's Theorem** If a graph has $n$ vertices and $n \geq 3$ and at least $n/2$ edges meet at each vertex, then the graph has a Hamiltonian circuit.

# A VARIETY OF VERSIONS AND USES OF INDUCTION

'*Proof*'  It is easy to check that the result holds for graphs with 3 vertices, so we have the basis for an inductive proof. To do the induction step, let $G$ be a graph on $n$ vertices which satisfies the stated condition on the edge number and which therefore, by our inductive hypothesis, has a Hamiltonian circuit.

Now make a new graph by adding an extra vertex $v$ and joining it to at least $\frac{1}{2}(n+1)$ of the other vertices. Consider any Hamiltonian circuit in $G$ (see Figure 5.5). There must be two adjacent vertices of this circuit joined to $v$. Why? [**5.38**] Now replace the edge $e$ of the circuit by edges $e'$ and $e''$ in Figure 5.5. The result is a Hamiltonian circuit in the new graph and the induction is complete.

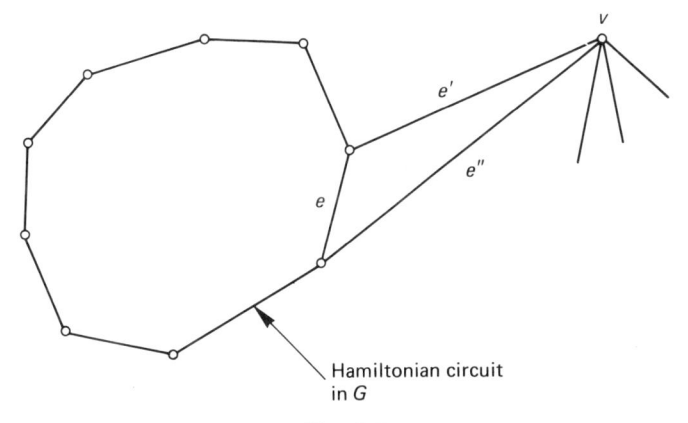

**Fig. 5.5**

[**5.39**] This time the result, Dirac's theorem, is correct, but what is wrong with our proof?

# 6 Permutations—in which ALICE is transformed

'But when... the great world-year has run out, the formation of a new world begins again, which is so exactly like the former one that in it all things, persons and phenomena, return exactly as before.'
Zeller, *Philosophie der Griechen*, Vol. III

This chapter departs from the main stream in having nothing ostensibly to do with numbers. So, if your main concern is understanding the number system, you lose nothing by skipping this chapter. If at some stage (and exactly which stage doesn't really matter) you tackle it, we hope you will gain from it an appreciation that many of the concepts discussed in the main stream with reference to numbers are really of much wider applicability. The concepts we have in mind are factorisation, uniqueness, well-defined and equivalence classes.

## 6.1 THE BASIC DEFINITION

First let us agree on what a permutation is. Suppose that we start with the set of letters of our heroine's name and arrange them in order A, L, I, C, E. If they are then rearranged (permuted) into L, I, E, C, A, we can describe this permutation, which we shall call $p$, for short, by the symbol

$$\begin{pmatrix} A & L & I & C & E \\ \downarrow & \downarrow & \downarrow & \downarrow & \downarrow \\ L & I & E & C & A \end{pmatrix}$$

where the arrows can be read as 'is replaced by'. Note that we are going to concentrate on what gets replaced by what, and not on the final arrangement, so that the symbol

$$\begin{pmatrix} A & L & E & C & I \\ L & I & A & C & E \end{pmatrix}$$

represents $p$ too. (The arrows are usually suppressed.) It just gives the same bits of information but in a different order. This, of course, should not surprise you, having got used to the idea in Chapter 3 that, for example, $\frac{3}{9}$ and $\frac{4}{12}$, are *different* symbols representing the *same* rational number.

A useful way of thinking of a permutation is as a function, or mapping, or operation. Just as the operation 'square' acting on 4 produces 16, so the operation $p$ acting on I produces E. A commonly used, and quite useful, piece of jargon is to refer to 16 as the *image* of 4 under 'square', and E as the image of I under $p$. This terminology is useful here in emphasising what distinguishes permutations from functions in general: suppose that we regard 'square' as acting on the real numbers in the closed interval $[-1, 2]$. We draw your attention to three important details.

(a) Some of the images lie outside this set—for example, $2^2 = 4 \notin [-1, 2]$.
(b) There exist *different* members of $[-1, 2]$ with the same image—for example, $1^2 = (-1)^2 = 1$.
(c) Some members of $[-1, 2]$ do not occur as images—for example, $-1$.

You are probably more familiar with the idea of a function than that of a permutation, so it makes sense to define a permutation somewhat negatively as a function not having any of the features (a), (b) and (c) above. Putting it more positively, we can say that a permutation is a function acting on a set $A$ such that all the images are members of $A$, no two members of $A$ have the same image, and every member of $A$ is the image of something in $A$. This is sometimes expressed by saying that a permutation is a one-to-one correspondence, or bijection, between $A$ and itself.

Note that 'square' becomes a permutation if we restrict it to act upon the closed interval $[0, 1]$—a fact neatly illustrated by the graphs of Figure 6.1.

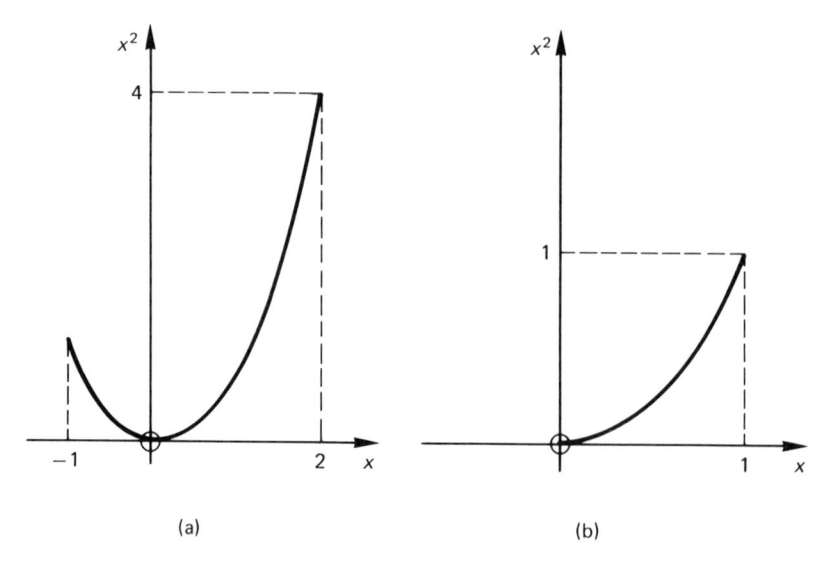

**Fig. 6.1**   (a) 'Square' acting on $[-1, 2]$; (b) 'Square' acting on $[0, 1]$

The function of Figure 6.1(b) is a permutation acting on an infinite set, whereas $p$ acts on the finite set $\{A, L, I, C, E\}$. From now onwards we shall restrict discussion to the latter type.

## 6.2 DO PERMUTATIONS BEHAVE LIKE NUMBERS?

Functions can be combined by performing them one-after-another, as in the expression log sin 30°. If you think about how this number would be worked out, it is clear that the sine function is applied first to 30°; then the logarithm is used to operate on the result.

This method of combining functions is called composition, and is denoted by ∘. Thus, $f \circ g$ is the function which is the result of applying $g$ first, then $f$, and is often read as '$f$ circle $g$'. [Beware—some authors reverse this order convention.]

[6.1] Show, from the definition of a permutation, that if $f$ and $g$ are permutations acting on the same set $A$, then so is $f \circ g$.

Using the method of representing permutations introduced in Section 6.1, we can, given the representations of $f$ and $g$, calculate the representation of $f \circ g$. An example will make the process clear. Suppose

$$f = \begin{pmatrix} A & L & I & C & E \\ I & E & C & A & L \end{pmatrix} \quad \text{and} \quad g = \begin{pmatrix} A & L & I & C & E \\ E & C & I & L & A \end{pmatrix}$$

$f \circ g$ is specified by giving the 'fates' of each of the letters, so, what happens to A? $g$ replaces it by E, which $f$ then replaces by L, so our description of $f \circ g$ begins

$$\begin{pmatrix} A & L & I & C & E \\ L & & & & \end{pmatrix}$$

Working out the fates of the other letters in the same way, we obtain

$$f \circ g = \begin{pmatrix} A & L & I & C & E \\ L & A & C & E & I \end{pmatrix}$$

A quicker way of doing this, which utilises the observation that the order in which the information is given in a permutation symbol is of no consequence, is the following scheme:

$$\begin{matrix} & A & L & I & C & E & \\ g \Big\{ & & & & & & \Big\} f \circ g \\ & E & C & I & L & A & \\ f \Big\{ & & & & & & \\ & L & A & C & E & I & \end{matrix}$$

If we write

$$\begin{pmatrix} A & L & I & C & E \\ L & A & C & E & I \end{pmatrix}$$

as the single permutation $h$, we have $h = f \circ g$, and comparing this with $10 = 5 \times 2$ suggests a rather tenuous analogy between arithmetic of natural numbers and the algebra of permutations in which permutations play the roles of natural numbers, and composition takes on the role of multiplication. Can the connection be made less tenuous and more fruitful? Is there a sense in which it may be helpful to think of $f \circ g$ as a factorisation of $h$? Well, perhaps, provided that we can think of reasonable analogues for prime numbers and for the particular number 1.

1 is easy. It is the unique number which has no effect under multiplication: $(\forall x)(1 \times x = x \times 1 = x)$. There is an obvious permutation on $\{A, L, I, C, E\}$ which has no effect under composition—namely,

$$\begin{pmatrix} A & L & I & C & E \\ A & L & I & C & E \end{pmatrix}$$

which we shall call $i$. Then, if $p$ is *any* other permutation on $\{A, L, I, C, E\}$, it is clear that $i \circ p = p \circ i = p$, and that $i$ is the only permutation with this property.

Now, what about primes? Prime numbers are the building blocks of $\mathbb{N}$, in the sense that any natural number except 1 is a product of primes. So, can we find a class of 'simple' permutations—'prime permutations', if you like—such that any permutation is expressible as a composition of primes?

One measure of simplicity of a permutation is the number of elements it 'moves'. For example,

$$\begin{pmatrix} A & L & I & C & E \\ C & L & I & A & E \end{pmatrix}$$

is simpler than

$$\begin{pmatrix} A & L & I & C & E \\ L & C & A & I & E \end{pmatrix}$$

because the former only moves two letters, A and C, whereas the latter moves four—all of them except E. The permutations which move just two elements are called transpositions, and, apart from $i$, they are the simplest. What about those which only move one element? [**6.2**]

Can we build all permutations from compositions of transpositions? Here is one way of doing it for one particular permutation.

**Example 1**

$$\underline{A}L\underline{I}CE \to LA\underline{I}CE \to LC\underline{I}AE \to LCAIE$$

so

$$\begin{pmatrix} A & L & I & C & E \\ L & C & A & I & E \end{pmatrix} =$$

$$\begin{pmatrix} A & L & I & C & E \\ I & L & A & C & E \end{pmatrix} \circ \begin{pmatrix} A & L & I & C & E \\ C & L & I & A & E \end{pmatrix} \circ \begin{pmatrix} A & L & I & C & E \\ L & A & I & C & E \end{pmatrix}$$

[**6.3**] Using this example as a guide, can you devise a systematic procedure (algorithm) for expressing any permutation on any number of symbols as a composition of transpositions?

If we take transpositions to be our 'primes', then we certainly don't get any *unique* prime factorisation theorem. Here are some other sequences of transpositions whose compositions all give the same result.

**Example 2**   $\underline{A}L\underline{I}CE \to LA\underline{I}CE \to LA\underline{C}IE \to LCAIE$, in which every transposition interchanges a consecutive pair.

**Example 3**   $\underline{A}L\underline{I}CE \to \underline{C}L\underline{I}AE \to \underline{L}C\underline{I}AE \to \underline{A}C\underline{I}LE \to \underline{I}CA\underline{L}E \to LCAIE$, in which every transposition moves the letter at the left-hand end.

**Example 4**   ALICE → AEICL → LEICA → LEIAC ↘
                LCAIE ← LCAEI ← LCIEA ← LCIAE ↗

in which they all involve the right-hand end.

**Example 5**   ALICE → ELICA → LEICA → LCIEA ↘
                       LCAIE ← LCEIA ↗

where E is moved at each step.

Clearly, there are many others.

So, if we can't have uniqueness, are the transpositions more like the $S$-primes of Chapter 2? Unfortunately, the situation is rather worse than this, and by studying the next two examples you should see why our hopes of building an analogy with primes and multiplication begin to crumble.

[6.4] $$\begin{pmatrix} A & L & I & C & E \\ L & A & I & C & E \end{pmatrix} = \begin{pmatrix} A & L & I & C & E \\ C & L & A & I & E \end{pmatrix} \circ \begin{pmatrix} A & L & I & C & E \\ L & I & C & A & E \end{pmatrix}$$

[6.5] Compare $$\begin{pmatrix} A & L & I & C & E \\ L & A & I & C & E \end{pmatrix} \circ \begin{pmatrix} A & L & I & C & E \\ A & I & L & C & E \end{pmatrix}$$

with

$$\begin{pmatrix} A & L & I & C & E \\ A & I & L & C & E \end{pmatrix} \circ \begin{pmatrix} A & L & I & C & E \\ L & A & I & C & E \end{pmatrix}$$

## 6.3 AN IMPORTANT INVARIANT

The analogy

$$\left\{ \begin{array}{c} \text{permutations} \rightarrow \text{numbers} \\ \text{composition} \rightarrow \text{multiplication} \\ i \rightarrow 1 \\ \text{transpositions} \rightarrow \text{primes} \end{array} \right\}$$

hasn't turned out to be very fruitful. But let us change track: instead of trying to dress up permutations in numerical clothes and being disappointed with the fit, we look at permutations on their own merits and ask how they *do* behave. We shall discover some nice algebra in the process.

Take another look at our transposition compositions which produced

$$\begin{pmatrix} A & L & I & C & E \\ L & C & A & I & E \end{pmatrix}$$

In all the examples, you will see that we used an odd number of transpositions, and this is no accident. We now state the general truth of which this is an example.

> **The Conservation of Parity** If $p$ is a permutation of $n$ objects which is expressible as the composite of an odd number of transpositions, then any other sequence of transpositions whose composition is $p$ will also have an odd number of transpositions. ($p$ is then called an *odd* permutation.) Similarly, if $q$ is achievable with an even number of transpositions, then it is *only* achievable with an even number of transpositions (and $q$ is called an *even* permutation).

Notice that this result enables us to classify each permutation of $n$ objects by its parity (oddness or evenness) in an unambiguous way. In other words, parity is a well-defined property of a permutation.

Our aim now is to account for this conservation law. We first need a result concerning the effect of a single transposition on a 'word'. Suppose that we start with ALICE and make a transposition (say, interchange A and C), to arrive at CLIAE. The letter pair {A, I} appears in the order AI in ALICE but IA in CLIAE, whereas {A, E} appears in the same order in both words. We now follow up this idea systematically. Below we have listed all the letter pairs from {A, L, I, C, E} (ten of them) in the same letter order in which they occur in ALICE. Underneath we give the same list, but each pair is written as it appears in CLIAE. The underlined ones are those which have suffered a switch.

|  |  |  |  |  |  |  |  |  |  |
|---|---|---|---|---|---|---|---|---|---|
| AL | AI | AC | AE | LI | LC | LE | IC | IE | CE |
| LA | IA | CA | AE | LI | CL | LE | CI | IE | CE |

Why are just these five pairs reversed? If we interchange A and C in ALICE, it is clear that the only pairs whose order is switched are those involving A and a letter between A and C, those involving C and a letter between A and C, and the pair AC itself. In the diagram below we have shown these five pairs.

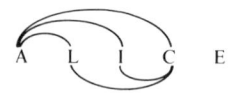

A L I C E

[**6.6**] By applying this line of argument, you should now be able to show that if we apply any transposition to a string of symbols $s_1 s_2 s_3 \ldots s_n$, the number of pairs which get switched is always odd.

[**6.7**] It is an easy step from this to see that if one string is converted to another by a sequence of transpositions, then the number of pairs which end up switched is odd or even, depending on whether the number of transpositions used is odd or even, respectively.

As an illustrative example consider the sequence

$$\text{ALICE} \rightarrow \text{LAICE} \rightarrow \text{EAICL} \rightarrow \text{ECIAL} \rightarrow \text{ECLAI}$$

Comparing ALICE with ECLAI, we see that *eight* pairs have been switched and *four* transpositions have been used—both even!

There is only one slight snag to spoil all this. To explain the worry, consider

$$\begin{pmatrix} A & L & I & C & E \\ E & C & L & A & I \end{pmatrix} \quad \text{and} \quad \begin{pmatrix} C & E & L & A & I \\ A & I & C & E & L \end{pmatrix}$$

These are two *representations* of the *same permutation*. All that you ([**6.7**]) have proved up to this point is that if the transition from ALICE to ECLAI is achieved by $n$ transpositions and results in $m$ pairs being switched, and if $n'$, $m'$, respectively, are the corresponding numbers for a transition from CELAI to AICEL, then $n$, $m$ have the same parity. What we do not know yet is whether $n$ and $n'$ (and, hence, $m$ and $m'$) have the same parity. If we lack a guarantee of this, we have only proved conservation of parity for *permutation representations* rather than for *permutations*.

Plugging this final hole depends on the following simple observation. Since

$$\begin{pmatrix} A & L & I & C & E \\ E & C & L & A & I \end{pmatrix} \quad \text{and} \quad \begin{pmatrix} C & E & L & A & I \\ A & I & C & E & L \end{pmatrix}$$

represent the same permutation, the first representation can be transformed into the second by a sequence of transpositions of whole columns. The process is illustrated below.

$$\begin{pmatrix} \boxed{A} & L & I & \boxed{C} & E \\ \boxed{E} & C & L & \boxed{A} & I \end{pmatrix} \to \begin{pmatrix} C & \boxed{L} & I & A & \boxed{E} \\ A & \boxed{C} & L & E & \boxed{I} \end{pmatrix} \to \begin{pmatrix} C & E & \boxed{I} & A & \boxed{L} \\ A & I & \boxed{L} & E & \boxed{C} \end{pmatrix} \to \begin{pmatrix} C & E & L & A & I \\ A & I & C & E & L \end{pmatrix}$$

Applying this sequence to the top rows only takes us from ALICE to CELAI. The same sequence applied to the bottom rows takes us from ECLAI to AICEL.

Putting all this together, suppose that $\binom{t}{b}$ and $\binom{t'}{b'}$ are representations of the same permutation ($t$ and $b$ stand for top and bottom rows), and let $k$ be the number of transpositions of columns used to get from $\binom{t}{b}$ to $\binom{t'}{b'}$. We obtain two routes from $t$ to $b'$, as shown in Figure 6.2. Hence, $t \to b'$ is achieved by $n + k$ transpositions by the sequence going via $b$, and by $k + n'$ transpositions by the sequence via $t'$.

So $n + k$ has the same parity as $k + n'$.

So $n$ has the same parity as $n'$, as was required, and our proof is complete. □

```
         k transpositions
  t ──────────────────▶ t'
  │                     │
  │                     │
n │                     │ n' transpositions
  transpositions        │
  ▼                     ▼
  b ──────────────────▶ b'
         k transpositions
```

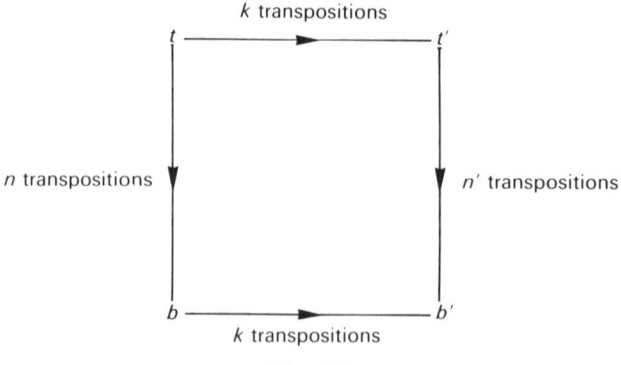

**Fig. 6.2**

## 6.4 HOW MANY ODDS?

We now know that there are just two types of permutations—the odd ones and the even ones. What we don't yet know is their distribution. Given $n$ letters, there are $n!$ permutations of them [**6.8**], but how many of them are even? With one trivial exception [**6.9**], whatever the value of $n$, exactly half the $n!$ permutations are even—and you couldn't really wish for a simpler answer than that! This *result* is important for permutation theory and for topics such as group theory, geometric symmetry and coding which make extensive use of permutations. The *method of proof* is important, because it finds application in many areas of mathematics having nothing to do with permutations. It is mainly for this latter reason that we have chosen to make a fuss of it here, but it fits neatly into this chapter, anyway, because it hinges on the idea of a bijection—the very idea which was used to define a permutation.

In outline the method is as follows. We have two finite sets, $A$ and $B$, which we want to show have the same number of elements. To do it, we associate with each element of $A$ a partner in $B$, in such a way that: (1) each element of $A$ has just one partner in $B$ and (2) each element of $B$ is the partner of exactly one element of $A$.

In Figure 6.3 we have shown various pairings which *do not* satisfy these conditions. (b) and (c) illustrate two ways in which (1) may fail; (a) and (d) show two ways in which (2) may fail.

A pairing between $A$ and $B$ which does satisfy (1) and (2) is called a bijection between $A$ and $B$, and it is now clear that if a bijection exists, then $A$ and $B$ must have the same number of elements.

So, to our use of this principle. Our sets $A$ and $B$ are the odd and even permutations, respectively, of the $n$ symbols, $s_1, s_2, \ldots, s_n$, and we wish to show that $A$ and $B$ are the same size.

Each permutation $a$ in $A$ can be expressed as the composition of an odd number of transpositions, so let $a = t_k \circ t_{k-1} \circ \cdots \circ t_2 \circ t_1$, where the $t_i$ are the

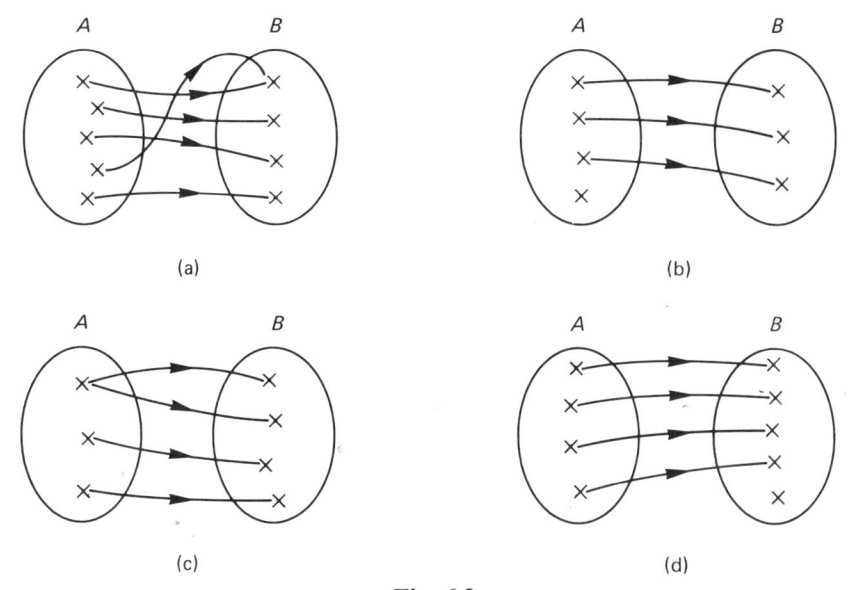

**Fig. 6.3**

transpositions and $k$ is odd. Now choose any fixed transposition $t$, say the switch of $s_1$ and $s_2$, and define $f(a) = t \circ a$.

For each $a \in A$, $f(a)$ is necessarily an even permutation, so it is in $B$. Thus, we have a pairing function $f$ from $A$ to $B$, and we need to show that it is a bijection. That is, that it satisfies properties (1) and (2). Now, our definition of $f$ specifies exactly the single partner of $a$ for each $a$ in $A$, so (1) is satisfied. Now we take any $b$ in $B$ and show that $b = f(a)$ for some $a$ in $A$. Since $b \in B$, it can be written as $b = t_m \circ t_{m-1} \circ \cdots \circ t_1$, where $m$ is even. But it can also be written as $b = t \circ t \circ t_m \circ t_{m-1} \circ \cdots \circ t_1$, since doing the same transposition twice in succession has no net effect. So we have $b = t \circ a = f(a)$, where $a = t \circ t_m \circ t_{m-1} \circ \cdots \circ t_1$, and since $m$ is even, we see that $a$ is an *odd* permutation. That is, $a \in A$. So, given any $b$ in $B$, we have found a partner for it in $A$. To complete the verification of (2), we have to show that $b$ has no other partner. Suppose that it did—say $a'$ was another one. Then we would have

$$f(a) = f(a') = b$$

which means

$$t \circ a = t \circ a' = b$$

Therefore,

$$t \circ t \circ a = t \circ t \circ a'$$

in other words,

$$i \circ a = i \circ a'$$

81

So

$$a = a'$$

and the proof is complete. □

## 6.5 UNIQUE FACTORISATION RESTORED

We have concentrated so far on analysing permutations into compositions of transpositions, for the good reason that transpositions were perceived as being, in a sense, simple.

Now consider the following five examples of permutations, each written in our standard symbolism and in another self-explanatory diagrammatic form.

(a) $\begin{pmatrix} A & L & I & C & E \\ L & I & C & E & A \end{pmatrix}$

(b) $\begin{pmatrix} A & L & I & C & E \\ C & E & A & L & I \end{pmatrix}$

(c) $\begin{pmatrix} A & L & I & C & E \\ A & E & C & L & I \end{pmatrix}$

(d) $\begin{pmatrix} A & L & I & C & E \\ C & I & L & E & A \end{pmatrix}$

(e) $\begin{pmatrix} L & O & B & A & T & C & H & E & V & S & K & Y \\ B & Y & V & L & K & E & T & C & S & A & H & O \end{pmatrix}$

The diagrams show how each one can be illustrated as a cycle or combination of cycles, and it is these cyclic permutations which we now want to concentrate on.

First, cycles admit a concise notation:

$$\begin{pmatrix} A & L & I & C & E \\ C & E & A & L & I \end{pmatrix}$$

for example, could be written as $(A \to C \to L \to E \to I \to A)$, which can be shortened by (A C L E I), with the understanding that the fate of each letter is the letter to the right of it, and that the whole symbol 'chases its tail'—I goes to A. (A C L E I) would be called a 5-cycle, and the notation is not unique just to the extent that (C L E I A), (L E I A C), (E I A C L) and (I A C L E) all represent the same permutation because of the tail-chasing.

If you look again at those examples which split into more than one cycle, you will see that the cycles are disjoint—no two of them have a letter in common. This is important, because although permutations in general do not commute (see [**6.5**]), if $p$ and $q$ are *disjoint cycles*, then $p \circ q = q \circ p$. [**6.10**]

These examples are not misleading: they are indications of our uniqueness theorem.

---

**Uniqueness Theorem**  Any permutation can be expressed uniquely as a composition of disjoint cycles.

---

It is a very easy result to prove (unlike the corresponding unique factorisation theorem of Chapter 2), and its proof is *constructive*—we obtain from it a *method of finding* the cycles.

*Proof*  Let $p$ be any permutation of $n$ 'letters', $s_1, s_2, \ldots s_n$. Pick any one of them (say $s_1$), then write down its fate (say $s_2$), then $s_2$'s fate $s_3, \ldots$, and so on until you reach a letter previously considered. Suppose that $s_k$ is the *first* letter to be repeated. Then $s_k = s_1$, for, if not, we would have the situation illustrated in Figure 6.4 in which $p$ would map $s_{k+l-1}$ to $s_k$ and $s_{k-1}$ to $s_k$. But $s_{k+l-1}$ and $s_{k-1}$ are different, so $p$ would not be a bijection (see condition (2) of Section 6.4), and therefore not a permutation.

$$s_1 \to s_2 \to s_3 \to \cdots \to s_{k-1} \to s_k \to s_{k+1} \to \cdots \to s_{k+l} = s_k$$

This cannot happen.

Fig. 6.4

Now the cycle $(s_1, s_2, \ldots, s_{k-1})$ specifies what $p$ does to each of $s_1$ to $s_{k-1}$. If $k - 1 = n$, $p$ is this single cycle. If not, pick a letter other than these $k-1$—say $t_1$—and generate another cycle in the same way $(t_1, t_2, \ldots, t_{l-1})$. These two cycles cannot have any letter in common. Why not? [**6.11**] Continuing in this way, $p$ is expressed as a composition of disjoint cycles.

[6.12] Use this process to obtain a disjoint cycle factorisation of

$$\begin{pmatrix} A & B & C & D & E & F & G & H & I & J & K & L & M & N \\ G & J & C & M & K & N & E & H & B & I & A & F & L & D \end{pmatrix}$$

The answer is unique apart from the order in which the cycles appear in the composition. To see this, suppose

$$p = c_1 \circ c_2 \circ c_3 \circ \cdots \circ c_k$$

and

$$p = d_1 \circ d_2 \circ d_3 \circ \cdots \circ d_l$$

where the $c_i$'s are disjoint cycles, as are the $d_i$'s, and $p$ acts on $\{s_1, s_2, \ldots, s_n\}$. Number the $c_i$'s and $d_i$'s so that $c_1$ and $d_1$ are the cycles involving $s_1$. Then $c_1$ and $d_1$ are identical cycles, for, if not, we could arrange both to start at $s_1$; then if the $i$th place was the first at which they differed, $c_1$ and $d_1$ would be saying that $p$ maps some element (the one in the $(i-1)$st place) onto two distinct elements (those in the $i$th places of $c$ and $d$), so again $p$ would not be a bijection.

[6.13] The disjointness condition is essential. If the 'factors' are not required to be disjoint, there are many ways of expressing any permutation as a composition of cycles.

## 6.6 ASSOCIATIVITY AND THE INVERSE OF A PERMUTATION

To every permutation $p$ there is an inverse permutation $p^{-1}$ which 'neutralises the effect of $p$'. We mean that if $p$ moves X to Y, $p^{-1}$ brings Y back to X, so that $p^{-1} \circ p$ has no effect. In other words, $p^{-1} \circ p = i$. An example will make this clear. If $p$ is

$$\begin{pmatrix} A & L & I & C & E \\ E & L & C & A & I \end{pmatrix}$$

then $p^{-1}$ is

$$\begin{pmatrix} A & L & I & C & E \\ C & L & E & I & A \end{pmatrix}$$

because $p$ moves C to A, so $p^{-1}$ must move A back to C, and so on for the other letters. A quicker way of getting $p^{-1}$ from $p$ is to interchange the two rows of $p$, so that

$$p^{-1} = \begin{pmatrix} E & L & C & A & I \\ A & L & I & C & E \end{pmatrix}$$

Obviously we also have $p \circ p^{-1} = i$, so any permutation commutes with its inverse.

Another important property of permutation composition is that it is *associative*. That is, $(p \circ q) \circ r = p \circ (q \circ r)$. We have implicitly assumed this in several places in this chapter (hoping you wouldn't notice!—or, more hopefully, expecting you to notice and check it for yourself!) To establish the property, we simply check that these two composites have the same effect on any 'letter', as follows.

$((p \circ q) \circ r)(A)$ means $(p \circ q)(r(A))$, which in turn means $p(q(r(A)))$. On the other hand, $(p \circ (q \circ r))(A)$ means $p((q \circ r)(A))$, which is $p(q(r(A)))$.

It now follows (see Chapter 5 for a rigorous inductive proof) that if we have a string of permutations to be composed, $p_1 \circ p_2 \circ p_3 \circ \cdots \circ p_k$, then, no matter how we bracket these in pairs, we always arrive at the same permutation. (Note that associativity holds for composition of functions in general, not just for permutations, and the method of proving this is identical.) So at least composition has the associative property in common with multiplication, even if commutativity is not shared.

Another faithful analogy is that between inverses of permutations and reciprocals of numbers:

$$p \circ p^{-1} = p^{-1} \circ p = i, \quad \text{just as} \quad 3 \times 3^{-1} = 3^{-1} \times 3 = 1$$

and

$$(p^{-1})^{-1} = p, \quad \text{just as} \quad (2^{-1})^{-1} = 2$$

**[6.14]** Show that $(p_1 \circ p_2 \circ \cdots \circ p_k)^{-1} = p_k^{-1} \circ p_{k-1}^{-1} \circ \cdots \circ p_2^{-1} \circ p_1^{-1}$.

**[6.15]** Every permutation has an inverse, but not every real number has a reciprocal. Which?

**[6.16]** There are just two real numbers which are self-reciprocal. Which? Can you characterise all permutations on $n$ letters which are self-inverse?

**[6.17]** If $p$ is the cycle $(a_1 a_2 a_3 \ldots a_m)$, what is the simplest way of expressing $p^{-1}$?

When $n$ is an integer and $x$ is any number, $x^n$ is defined by $x \times x \times x \cdots \times x$ ($n$ factors) if $n$ is positive, by 1 if $n$ is zero and by $1/x^{-n}$ if $n$ is negative. Similarly, if $p$ is a permutation, it is consistent to define $p^n$ as $p \circ p \circ p \circ \cdots \circ p$ ($n$ 'factors') if $n$ is a positive integer, as $i$ if $n$ is zero and as $(p^{-n})^{-1}$ if $n$ is a negative integer.

**[6.18]** If $p$ and $q$ are permutations and $m$, $n$ are integers, which of the following hold: (i) $p^m \circ p^n = p^{m+n}$; (ii) $(p^m)^n = p^{mn}$; (iii) $(p \circ q)^n = p^n \circ q^n$?

**[6.19]** If $p$ is any permutation on a finite number of letters, there must be some natural number $n$ for which $p^n = i$. Why is this so?

**[6.20]** The smallest natural number $n$ for which $p^n = i$ is called the order of $p$. If $p$ is a cycle $(a_1 \, a_2 \ldots a_k)$, and we call $k$ the length of the cycle, what is the connection between the order of $p$ and its length?

**[6.21]** If $p$ is written in *disjoint cycle* notation as $c_1 \circ c_2 \circ \cdots \circ c_l$, what is the connection between the order of $p$ and the lengths of the $c_i$'s? [*Hint*: Do some specific cases to get some data to work on.]

**[6.22]** If $p$ is a cycle, what is the connection between its length and its parity?

## 6.7 A MORE FRUITFUL NUMERICAL ANALOGY

There is something preventing our analogy between powers of permutations and powers of numbers from being as satisfying as we might wish. In the symbol $x^n$ in 'ordinary' algebra, the base $x$ and the exponent $n$ are both numbers. In $p^n$, $p$ is a permutation but $n$ is still a number. Wouldn't it be nice if we could think of some way of interpreting $p^q$ when $p$ and $q$ are *both* permutations.

Let us make the apparently perverse definition that $p^q$, called the conjugate of $p$ by $q$, means $q \circ p \circ q^{-1}$.

**[6.23]** Check that with this definition, if we 'translate' numbers into permutations, multiplication into composition, 1 into $i$, reciprocals into inverses and powers into conjugates, the following all hold: (i) $(xy)^z = x^z y^z$; (ii) $(x^y)^z = x^{zy}$ (but *not* $(x^y)^z = x^{yz}$); (iii) $(1/x)^y = 1/x^y$; (iv) $x^1 = x$.

The significance of conjugation of permutations goes far beyond proving a nice analogy with ordinary algebra. Its real use is in helping us to analyse the structure of the set of $n!$ permutations of $n$ letters. To make a start on this analysis, we first make a connection with Chapter 3.

**[6.24]** Show that if we define $p \sim q$ to mean $p = q^r$ for some permutation $r$, then $\sim$ is an equivalence relation.

What do the equivalence classes look like? To answer this, we first observe that, given $q$ and $r$, there is an easy way of seeing what $q^r$ does without tracing the fates of every one of the $n$ letters through $r \circ q \circ r^{-1}$.

**[6.25]** Show that if $q$ sends A to B then $q^r$ sends $r(A)$ to $r(B)$.

For an example of this result, take $q$ to be

$$\begin{pmatrix} A & L & I & C & E \\ L & I & C & E & A \end{pmatrix}$$

and $r$ to be

$$\begin{pmatrix} A & L & I & C & E \\ E & L & C & I & A \end{pmatrix}$$

Then $q^r$ is

$$\begin{pmatrix} r(A) & r(L) & r(I) & r(C) & r(E) \\ r(L) & r(I) & r(C) & r(E) & r(A) \end{pmatrix} = \begin{pmatrix} E & L & C & I & A \\ L & C & I & A & E \end{pmatrix}$$

The significance of this is clearer if we do it using the disjoint cycle representation. For example, if we conjugate (ABC)∘(DE)∘(FGHI)∘(JK) by the permutation $r$, we get

$$(r(A)r(B)r(C))\circ(r(D)r(E))\circ(r(F)r(G)r(H)r(I))\circ(r(J)r(K))$$

which is permutation with the *same disjoint cycle structure* as the original—namely, the composite of two cycles of length 2, one of length 3 and one of length 4.

So, if $q$ and $r$ are any two permutations, $q$ and $q^r$ have the same disjoint cycle structure.

The converse is also true: if $p$ and $q$ have the same cycle structure, then $p$ is $q^r$ for some permutation $r$. This is proved by actually finding a suitable $r$. For example, if $p$ is

$$(ABC)\circ(DEF)\circ(GH)\circ(IJ)\circ(KL)\circ(M)\circ(N)$$

and $q$ is

$$(AMG)\circ(BIJ)\circ(KN)\circ(CL)\circ(DF)\circ(E)\circ(H)$$

all we need is an $r$ for which $r(A) = A$, $r(M) = B$, $r(G) = C$, $r(B) = D,\ldots$, etc. So the permutation we want is

$$r = \begin{pmatrix} A & M & G & B & I & J & K & N & C & L & D & F & E & H \\ A & B & C & D & E & F & G & H & I & J & K & L & M & N \end{pmatrix}$$

which in cycle notation is

$$(A)\circ(MBDKGCIE)\circ(FLJ)\circ(HN).$$

$r$ is *not* unique, nor is its cycle structure determined by the requirement that $p = q^r$. To see this, suppose that we write $q$ in the alternative form,

$$(MGA)\circ(BIJ)\circ(LC)\circ(KN)\circ(DF)\circ(H)\circ(E)$$

and construct $r$ in the same way as before. This time we get

$$r = \begin{pmatrix} M & G & A & B & I & J & L & C & K & N & D & F & H & E \\ A & B & C & D & E & F & G & H & I & J & K & L & M & N \end{pmatrix}$$

whose cycle structure is

$$r = (MACH)\circ(GBDKIENJFL)$$

To summarise then, $p$ and $q$ are conjugate permutations if and only if they have the same disjoint cycle structure. Given conjugate $p$ and $q$, the $r$ which makes $p = q^r$ is not unique—and we have not been able to say anything about its structure.

As a final exercise:

[6.26] Consider the set $\Omega$ of all permutations on $\{A, L, I, C, E\}$. There are $5! = 120$ of them. How many equivalence classes are there, and how many permutations are there in each of the classes? (Remember that you have a check: the total number in all the classes must be 120—and don't assume that all the equivalence classes are the same size, because they're not! What is the smallest class?)

If you would like to learn more about permutations—in particular, their connection with the important topics of symmetry and groups, and some of their applications, a good start would be *Permutations and Groups*, by C. D. H. Cooper. But be warned: he uses the opposite convention from ours, so that his *pq* means $q \circ p$, not $p \circ q$, and this affects some of the details.

# 7  Nests—in which the rationals give birth to the reals and the scene is set for arithmetic in ℝ

'All changed, changed utterly: A terrible beauty is born.'

W. B. Yeats

## 7.1 NESTS IN THE FOREST

Deep in conversation, Alice and the Tweedle twins have wandered into an unfamiliar part of the forest.

DEE: But there can't be any numbers which are not rational!

DUM: I'm inclined to agree, but can we prove it?

DEE: I've had a brilliant idea! Let $x$ be a number; then either it's an integer and, hence, it's a rational, or else there exists an integer $n$ with $n < x < n+1$. Now, if $x$ is not a rational, then either it lies between $n$ and $n + \frac{1}{2}$ or else it lies between $n + \frac{1}{2}$ and $n+1$.

DUM: So we're half-way there! Get it?

DEE: Be serious, Dum—at the next stage $x$ will lie between two consecutive rationals in the list $n, n + \frac{1}{4}, n + \frac{1}{2}, n + \frac{3}{4}, n + 1$. If we carry on like this, we can always trap $x$ between two rationals whose distance apart is less than $\frac{1}{2^n}$ for any natural number $n$.

DUM: That still doesn't show that $x$ is a rational—only that it lies between pairs of rationals that are arbitrarily close together. Can you help us, Alice?

ALICE: Not really; but I've been thinking.... The square root of 4 is 2 and the square root of 9 is 3, so the square root of 5 lies between 2 and 3. Since $(2\frac{1}{2})^2 = 6\frac{1}{4}$, then $\sqrt{5}$ lies between 2 and $2\frac{1}{2}$. And since $(2\frac{1}{4})^2 = 5\frac{1}{16}$, then $\sqrt{5}$ lies between 2 and $2\frac{1}{4}$, and so on... but $\sqrt{5}$ is not a rational!

DEE AND DUM (*in unison*): Prove it!

ALICE: OK. Suppose that $\sqrt{5} = 2 + p/q$, where $p$ and $q$ are integers and $p$ is the least possible integer numerator. Clearly, $0 < p < q$. Now a bit of school algebra gives us the equation $q^2 - 4pq = p^2$. In other words, $q$ divides $p^2$, so we can write $p^2 = qr$ for some integer $r$. Now for the clever bit!

$$\sqrt{5} = 2 + p^2/pq = 2 + qr/pq = 2 + r/p$$

where $r/p$ is a rational whose numerator $r$ is smaller than $p$, the numerator of $p/q$. There's your contradiction! Hence, $\sqrt{5}$ is not a rational, even though it can be approximated as closely as you like by a rational.

DUM: Um, I suppose you're correct. But $\sqrt{5}$ is sort of defined by successively closer pairs of rationals. Perhaps that blackboard over there is relevant.

The three adventurers turn and observe a blackboard on which is written the following:

---
A real number is an equivalence class of rational nests.

---

Various bits of paper are flying about, which Dee and Dum collect excitedly. From these they are able to piece together the following pertinent definitions.

(1) A *closed rational interval* is a set of the form
$$I_n = \{x : x \in \mathbb{Q} \text{ and } p_n \leq x \leq q_n, \text{ where } p_n \in \mathbb{Q} \text{ and } q_n \in \mathbb{Q}\}$$
$$= [p_n, q_n]_\mathbb{Q}$$

(2) A *rational nest* is a collection $\{I_n\}$ of closed rational intervals satisfying

(i) $I_{n+1} \subseteq I_n \forall n \in \mathbb{N}$

and

(ii) for each $n \in \mathbb{N}$, $\exists m \in \mathbb{N}$ such that $|I_m| \leq \dfrac{1}{2^n}$

(3) Rational nests $\{I_n\}$ and $\{J_n\}$ are *equivalent* if and only if
$$I_n \cap J_n \neq \emptyset, \quad \forall n \in \mathbb{N}$$

ALICE: This is great. We had a rational nest for $\sqrt{5}$, remember.
$$I_1 = [2, 3]_\mathbb{Q}, \quad I_2 = [2, 2\tfrac{1}{2}]_\mathbb{Q}, \quad I_3 = [2, 2\tfrac{1}{4}]_\mathbb{Q}$$
and if $I_n = [p_n, q_n]$, then
$$I_{n+1} = \begin{cases} [p_n, r_n] & \text{when } p_n^2 \leq 5 \leq r_n^2 \\ [r_n, q_n] & \text{when } r_n^2 < 5 \leq q_n^2 \end{cases}$$
where $r_n = \tfrac{1}{2}(p_n + q_n)$, the mid-point of $I_n$. Thus (i) is satisfied by my construction, and notice that $|I_{n+1}| = \tfrac{1}{2}|I_n|$ and $|I_1| = 1$, so that $|I_n| = \dfrac{1}{2^n}$, which clinches (ii).

DEE: I can think of a rational nest defining any rational $r$ you like. How about
$$I_n = \left[r - \dfrac{1}{2^{n+1}}, r + \dfrac{1}{2^{n+1}}\right]_\mathbb{Q}$$

DUM: I've got a simpler one—namely, $J_n = \{r\}$ or, if you prefer, $J_n = [r, r]_\mathbb{Q}$.
ALICE (*excitedly*): That's why the equivalence of nests is there of course. Don't you see, Dum and Dee? We need to be able to recognise when different rational nests define the same number. Since that number must lie in every $I_n$ of a given defining nest $\{I_n\}$, and in every $J_n$ of some alternative defining nest $\{J_n\}$, we need $I_n \cap J_n \neq \emptyset$ for every $n \in \mathbb{N}$.
DEE: But is it an equivalence relation in the usual sense? I can't see any bits of paper around with it written down. We'll just have to prove it for ourselves. Let's write $\{I_n\} \sim \{J_n\}$ whenever $I_n \cap J_n \neq \emptyset$, $\forall n \in \mathbb{N}$. Obviously $\{I_n\} \sim \{I_n\}$ for all $n$, and if $\{I_n\} \sim \{J_n\}$, then $\{J_n\} \sim \{I_n\}$.
DUM: That's just using the commutative law for intersections of sets.
ALICE: OK, Dum, stop showing off! So, boys, $\sim$ is reflexive and symmetric, but what about transitivity?
DUM: Alice, which is better, complete happiness or a cheese sandwich?
ALICE: Complete happiness, of course, but what's that got to do with our problem?
DUM: Wrong! A cheese sandwich is better, 'cos nothing is better than complete happiness, and a cheese sandwich is better than nothing.
ALICE: What are you eating, Dee?
DEE: Some cheese sandwiches I found over there by that tree.
ALICE: OK. Let's have a working lunch. Suppose that we have rational nests $\{I_n\}$, $\{J_n\}$ and $\{K_n\}$ such that $I_n \cap J_n \neq \emptyset$ and $J_n \cap K_n \neq \emptyset$, $\forall n \in \mathbb{N}$. I wonder what would happen if for some particular $s$ we had $I_s \cap K_s = \emptyset$?
DEE: Well, then there would be a rational 'distance'—$L$, say—between $I_s$ and $K_s$, as in my picture.

Dee, having found a pencil, has drawn the following picture on his table napkin.

```
•————————•  Is              •————————•  Ks
       <------L------>
```

DEE (*continues with his mouth full*): Also, there is some $n \in \mathbb{N}$, such that $\dfrac{1}{2^n} < L$.

Since $\{J_n\}$ is a rational nest, there must be some $m \in \mathbb{N}$, $m > s$ with $|J_m| \leq \dfrac{1}{2^n}$. Now, $J_m$ cannot meet both $I_m$ and $K_m$, since its length is less than $L$ and the gap between $I_m$ and $K_m$ must be more than $L$. See...

```
            •————•   Jm
    •————————•   Im                        •————•  Km
•—————————————•  Is             •————————————•  Ks
```

So it's impossible for $I_s$ not to meet $K_s$!
ALICE: Brilliant, Dee! You've shown that $\{I_n\} \sim \{J_n\}$ and $\{J_n\} \sim \{K_n\}$ imply that $\{I_n\} \sim \{K_n\}$, so now we have transitivity. Have another cheese sandwich.
DUM: So now we have a criterion for telling when two rational nests define

the same number, and so, as it says on the blackboard, real numbers are just equivalence classes of rational nests. I'm tired; let's go to sleep.

## 7.2 ARITHMETIC WITH NESTS

The following day Alice seems troubled. Dee and Dum are happily constructing a rational nest defining $2\sqrt{5}$.

DUM: Come on, Alice, we've got as far as $I_6 = [4\frac{15}{32}, 4\frac{31}{64}]_\mathbb{Q}$.
ALICE: I'm worried about how we can do arithmetic with these real numbers like we do with rationals. I can't even see how to begin.
DEE: Well, if we could 'add' rational nests, then perhaps we could 'add' equivalence classes of rational nests—so long as our 'addition' is well defined. And perhaps 'adding' rational nests follows from 'adding' closed rational intervals. But how can we add two sets?
DUM: Or multiply them!?
ALICE: Let's invent Arithmetic on Intervals... before lunch.

By the time the sun is high in the sky, Alice and the twins are convinced that they have invented Arithmetic on Intervals. They are surrounded by screwed-up pieces of paper containing their half-attempts, their checkings, and so on. On the blackboard Alice has neatly summarised their conclusions.

The following are closed rational intervals:

$$I + J = \{x + y : x \in I \text{ and } y \in J\}$$
$$I \times J = \{x \cdot y : x \in I \text{ and } y \in J\}$$
$$-I = \{-x : x \in I\}$$
$$1/I = \{1/x : x \in I \text{ and } x \neq 0\}$$

where $I$ and $J$ are closed rational intervals.

But are the trio correct? Convince yourself that $I + J$, $I \times J$ and $-I$ are all closed rational intervals. **[7.1]** Show further that $1/I$ is also a closed rational interval, provided that $I$ does not contain 0. **[7.2]** We can forgive them that error! Show that $[p, q]_\mathbb{Q} + [p', q']_\mathbb{Q}$ is equal to $[p + p', q + q']_\mathbb{Q}$. **[7.3]** Find a simple formula for $[p, q]_\mathbb{Q} \times [p', q']_\mathbb{Q}$ when $p$ and $p'$ are both positive rationals. **[7.4]**

DEE (*looking at the blackboard*): The last one's wrong, Alice... $1/[-1, 2] = \{x : x \in \mathbb{Q} \text{ and either } x \leq -1 \text{ or } x \geq \frac{1}{2}\}$ isn't a closed interval.
ALICE: Um! Perhaps we'd better ban $1/I$ whenever $0 \in I$. That's a bit like not allowing $1/0$ with integer arithmetic.
DUM: You win some, you lose some. That's life!
ALICE: OK, you two; now for Arithmetic of Rational Nests. If $\{I_n\}$ and $\{J_n\}$ are two rational nests, then I bet that $\{I_n + J_n\}$ and $\{I_n \times J_n\}$ are

rational nests as well. If you two check that, I'll see whether or not the following are well-defined operations on rational nests.
$$\{I_n\} + \{J_n\} = \{I_n + J_n\}$$
$$\{I_n\} \times \{J_n\} = \{I_n \times J_n\}$$

Alice, Dee and Dum spend the warm afternoon in the forest feverishly checking that they have indeed defined addition and multiplication of real numbers. For addition they are correct, as we shall shortly see. Verify the salient facts for multiplication for yourself. [7.5]

---

If $\{I_n\}$ and $\{J_n\}$ are rational nests, then so is $\{I_n + J_n\}$.

---

We already know that $I_n + J_n$ is a closed rational interval for each $n \in \mathbb{N}$ by the earlier Arithmetic on Intervals. A typical element of $I_{n+1} + J_{n+1}$ is $x + y$, where $x \in I_{n+1}$ and $y \in J_{n+1}$. But $\{I_n\}$ is a nest, so $x \in I_n$. Similarly, $y \in J_n$. Hence, $x + y \in I_n + J_n$. In other words, $I_{n+1} + J_{n+1} \subseteq I_n + J_n$, which establishes criterion (i) (page 90) for a rational nest.

Now for any $n \in \mathbb{N}$, $\exists m \in \mathbb{N}$ such that $|I_m| \leq \frac{1}{2^{n+1}}$ and $|J_m| < \frac{1}{2^{n+1}}$. If $I_m = [p_m, q_m]_\mathbb{Q}$ and $J_m = [p_m', q_m']_\mathbb{Q}$, then $I_m + J_m = [p_m + p_m', q_m + q_m']_\mathbb{Q}$ and so $|I_m + J_m| = (q_m' - p_m') + (q_m - p_m) = |I_m| + |J_m| \leq \frac{1}{2^{n+1}} + \frac{1}{2^{n+1}} = \frac{1}{2^n}$. In other words, criterion (ii) (page 90) for a rational nest is fulfilled.

---

$\{I_n\} + \{J_n\} = \{I_n + J_n\}$ is well-defined.

---

Suppose that $\{I_n\} \sim \{I_n'\}$ and $\{J_n\} \sim \{J_n'\}$. By the definition of equivalence of rational nests, for each $n \in \mathbb{N}$, $\exists x \in \mathbb{Q}$ and $y \in \mathbb{Q}$ such that $x \in I_n \cap I_n'$ and $y \in J_n \cap J_n'$. But then $x + y$ is an element of both $I_n + J_n$ and $I_n' + J_n'$. In other words, $\{I_n + J_n\} \sim \{I_n' + J_n'\}$.

After supper the three plan the following day's activities.

ALICE: I can see that it will be easy to show that $-\{I_n\} = \{-I_n\}$ will define a rational nest. And then subtraction of real numbers will follow via
$$\{I_n\} - \{J_n\} = \{I_n\} + (-\{J_n\}) = \{I_n\} + \{-J_n\} = \{I_n - J_n\}$$

DEE: Division's going to be a problem because of our earlier difficulty with $1/I$ only being defined when 0 is not in $I$.
DUM: That means we cannot divide by a real number when any of the closed rational intervals in its defining nest contain zero.
ALICE: Leave them out!
DUM: What do you mean?

ALICE: Well, suppose that we have a nest $\{I_n\}$ defining a real number $x$ which is not zero. Clearly, 0 cannot lie in all the $I_n$'s, so there is an $m \in \mathbb{N}$ with 0 not in $I_m$. Hence, 0 is not in any of the subsequent $I_n$'s. Now we just define a new nest $\{J_n\}$ by $J_1 = I_m$, $J_2 = I_{m+1}$, etc. Our new nest is automatically equivalent to the original nest.

DEE: So $\{J_n\}$ defines $x$ as well and we can have $1/J_n$ for each interval in the nest... oh, wonderful! I shall sleep well tonight. Goodnight.

ALICE and DUM: Goodnight.

While our friends are asleep, show that every non-zero real number has a reciprocal. That is, show that $1/\{I_n\} = \{1/I_n\}$ is a rational nest, where, of course, $\{I_n\}$ is a suitable rational nest defining a non-zero real number and verify that $\{I_n\} \times \{1/I_n\}$ is a rational nest defining the number 1. [7.6] Finally, provide the details necessary to well-define division of real numbers. [7.7]

## 7.3 THE COMPLETE PICTURE

It is raining, and overnight a strange obelisk has appeared in the clearing of the forest occupied by Alice, Dee and Dum. What appear like ancient writings adorn the sides of the obelisk, mostly worn away by centuries of weather.

DUM: Hey, look at this, you two... can you read any of it?

ALICE: There's a bit here that says 'from the book *Alice in Numberland*, by John Baylis and Rod Haggarty'.

DEE: I've found a bit here... 'Chapter 8: Axioms for $\mathbb{R}$'... I can't read it... further down it says 'the *axioms of arithmetic* A1 $a + ...$' can't read any more... 'A3 $\exists$ a unique element $0 \in \mathbb{R}$ such that $a + 0 = a$'... some bits broken off here... 'A9 $a \cdot (b + c) = a \cdot b + a \cdot c$'.

DUM: Sounds like a list of obvious properties of rationals to me... but what is $\mathbb{R}$?

ALICE: Real numbers... it's a list of elementary properties of real numbers. If we've really constructed them, then they should satisfy these properties. A pity we can't see them all.

You, the reader, can, of course, see them all—turn to Section 8.1, where the Axioms of Arithmetic are listed in full. Every single one of them can be established for equivalence classes of rational nests (our real numbers) just by using the corresponding properties of rational numbers. [7.8] Why not turn to Chapter 8 now? Read Section 8.2 and then return to us here! Dee will have established A9 for you in the meantime.

DEE: Let's take rational nests $\{I_n\}$, $\{J_n\}$ and $\{K_n\}$ defining the real numbers $a$, $b$ and $c$, respectively. Now

$$\{I_n\} \times (\{J_n\} + \{K_n\}) = \{I_n\} \times \{J_n + K_n\}$$
$$= \{I_n \times (J_n + K_n)\} = \{L_n\}$$

and

$$(\{I_n\} \times \{J_n\}) + (\{I_n\} \times \{K_n\}) = \{I_n \times J_n\} + \{I_n \times K_n\}$$
$$= \{(I_n \times J_n) + (I_n \times K_n)\} = \{L_n'\}$$

Now a rational in $I_n \times (J_n + K_n)$ is of the form $x \cdot (y + z)$, where $x$, $y$ and $z$ are rationals in $I_n$, $J_n$ and $K_n$, respectively. But $x \cdot (y + z) = x \cdot y + x \cdot z$ is true for rational numbers and $x \cdot y + x \cdot z$ is an element of $(I_n \times J_n) + (I_n \times K_n)$. Hence, we have that $\{L_n\} \sim \{L_n'\}$. That proves that $a \cdot (b + c) = a \cdot b + a \cdot c$ for real numbers. Hurray!

DUM: But where did this obelisk come from? I've never seen all this sort of mathematics written down before!

ALICE: I have, in Chapter 1 of the Analysis book I used at university—what was it called? Ah, yes, *A First Elementary Introductory Course in Mathematical Analysis*. I didn't understand a word of it!

DEE: If we had a copy of the book the obelisk mentions, then we could check all the axioms of arithmetic from A1 to A9 and that would be fun.

ALICE: There's more axioms over this side... '*The axioms of order*'. I can't read any more.

DEE: We haven't ordered our real numbers yet.

DUM: You mean, tell them what to do?

DEE: No, silly! We need to know when $a \leqslant b$ and check that all the usual properties of an order relation hold.

DUM: That sounds like more hard work to me, and we've run out of cheese sandwiches!

ALICE: Come on, let's do it! If we take two real numbers $a$ and $b$ defined by rational nests $\{I_n\}$ and $\{J_n\}$, respectively, then since $a$ and $b$ are distinct, we must have $I_m \cap J_m = \varnothing$ for some $m \in \mathbb{N}$. So $I_m$ and $J_m$ must look like this

$$\bullet\!\!-\!\!-\!\!-\!\!-\!\!-\!\!-\!\!\bullet \; I_m \qquad\qquad \bullet\!\!-\!\!-\!\!-\!\!-\!\!-\!\!-\!\!\bullet \; J_m$$

or this

$$\bullet\!\!-\!\!-\!\!-\!\!-\!\!-\!\!-\!\!\bullet \; J_m \qquad\qquad \bullet\!\!-\!\!-\!\!-\!\!-\!\!-\!\!-\!\!\bullet \; I_m$$

In the former case $a < b$ and in the second case $b < a$. Then I'm pretty sure that the order properties of the rational numbers carry over to the real numbers.

Now read Section 8.3 and verify that Alice, Dee and Dum's real numbers satisfy A10–A14. [**7.9**]

DUM: I'm getting confused again—we've used rational nests to define real numbers some of which are not rational numbers. But this obelisk seems to show that the real numbers have exactly the same properties as the rationals—so what's $\mathbb{R}$ got that $\mathbb{Q}$ hasn't?

ALICE: It's written in gold letters at the top of the obelisk.

ALICE IN NUMBERLAND

> The real numbers are complete, since every non-empty subset of $\mathbb{R}$ which is bounded above has a smallest upper bound.

DUM: I'm completely lost.

DEE: Look, Dum: the non-rational real numbers, the irrationals, are like 'gaps' between rationals but not gaps in the usual sense... oh, this is hard to explain... but one could imagine a set of rationals like $S = \{x \in \mathbb{Q} : x^2 \leq 5\}$, where elements of $S$ are all clearly less than 3, but there is no smallest rational bigger than all the elements of $S$ because $\sqrt{5}$ is the smallest number bigger than all the elements of $S$ and $\sqrt{5}$ is irrational. So (*taking a big breath*), the irrationals sort of complete the picture!

DUM: Well, I can see what you mean with $\sqrt{5}$ and $S$, but that completeness axiom on the obelisk has got to work for any non-empty set $S$ of real numbers which is bounded above.

ALICE: Well, boys—here we go again! It's getting dark, so I hope you won't mind if I use the order relation on real numbers whenever I feel like it without having to justify every step. So let $S$ be a non-empty set of real numbers bounded above by $B_0$, say. Choose $B_1$ to be the right-hand end-point of one of the rational intervals $I_n$ in a rational nest $\{I_n\}$ defining $B_0$. Now $B_1$ is a rational upper bound for $S$. Let $s \in S$ and choose $b_1$ to be the left-hand end-point of some rational interval $J_n$ in a rational nest $\{J_n\}$ which defines $s$. Now $K_1 = [b_1, B_1]_\mathbb{Q}$ is a closed rational interval containing elements of $S$ whose right-hand end-point is an upper bound for $S$.

DUM: Help! I'm lost.

DEE: This picture will help.

On the base of the obelisk Dee has chalked the following Mathematical Graffito.

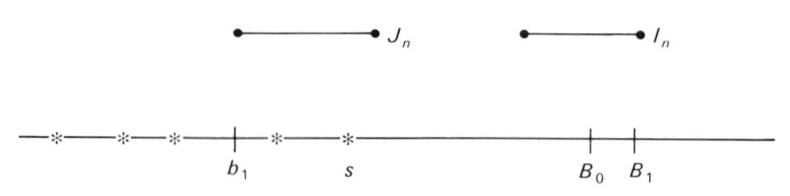

$$\bullet\!\!-\!\!-\!\!-\!\!-\!\!-\!\!-\!\!-\!\!\bullet\, J_n \qquad \bullet\!\!-\!\!-\!\!-\!\!-\!\!-\!\!-\!\!\bullet\, I_n$$

$$\text{---}*\text{---}*\text{---}*\text{---}|\text{---}*\text{---}*\text{-----------------}|\text{---}|\text{---}$$
$$\qquad\qquad b_1 \qquad\quad s \qquad\qquad\qquad\qquad B_0\ B_1$$

* denotes elements of $S$

ALICE: Thank you, Dee. OK. Here's my strategy—I'm going to trap the smallest upper bound for $S$ by constructing a rational nest which defines the supremum; I think that's what it's called; $K_1$ is the first rational interval in my nest. Now for $K_2$. Consider the rational midpoint of $K_1$—namely, $\frac{1}{2}(b_1 + B_1)$. Either it is an upper bound for $S$, in which case I shall take $K_2$ to be the left-hand half of $K_1$, or else it is not and so the right-hand half of $K_1$ contains elements of $S$, in which case that will be $K_2$. Whichever half it is, define the rationals $b_2$ and $B_2$ by setting $K_2 = [b_2, B_2]_\mathbb{Q}$. Carrying on in this way, we can define a rational nest $\{K_n\}$ where if $K_n = [b_n, B_n]_\mathbb{Q}$, then $K_{n+1} = [b_n, \frac{1}{2}(b_n + B_n)]_\mathbb{Q}$ or $[\frac{1}{2}(b_n + B_n), B_n]_\mathbb{Q}$ according as $\frac{1}{2}(b_n + B_n)$ is or is not an upper bound for $S$. The real number $B$ defined by the nest $\{K_n\}$ is then the supremum of $S$!

DEE: Gee, that's great, Alice, but is it clear that $B$ *is* an upper bound and can we prove that it is the *smallest* upper bound?

DUM: You two can sort that out. My head's spinning; I'm off to bed. Goodnight!

DEE: They really do exist, these real numbers, don't they, Alice?

ALICE: Yes, Dee, given that all numbers, even 1, 2 and 3, are all abstractions anyway. Look, Dee, I'm pretty tired myself. Let's go to sleep and leave the final details of the completeness of the real numbers to someone else.

That someone else is, of course, us—we prove first that the $B$ manufactured by Alice is indeed an upper bound for $S$. We proceed by contradiction. If $B$ is not an upper bound for $S$, then $\exists s \in S$ such that $s > B$. Now, this $s$ lies in every $K_n$, since the right-hand end-point of each $K_n$ is an upper bound for $S$. If $\{L_n\}$ is a rational nest defining $s$, then $L_n \cap K_n \neq \emptyset$ for all natural numbers $n$. Hence, $\{L_n\} \sim \{K_n\}$ and so $s = B$. This contradicts our original assumption that $s > B$. Hence, that assumption must be wrong, $B$ is an upper bound for $S$, as required. Finally, we show that $B$ is the smallest possible upper bound for $S$, again by contradiction. If $B'$ is a smaller upper bound for $S$, then $B' < B$. But every $K_n$ in the defining nest for $B$ contains $B$ and elements of $S$. Hence, $B'$ is contained in every $K_n$ and so $B'$ must equal $B$! This is our desired contradiction. Notice that it was wise of our three friends to retire for the day, because these last two proofs by contradiction may have proved too confusing, even for Alice. In each of them we have denied the result we want to prove and have then essentially deduced that very result. Thus, our original denial is false and our desired conclusion is true! GOODNIGHT!

# 8 Axioms for ℝ—in which we invent Arithmetic, Order our numbers and Complete our description of the reals

'In midair somewhere he lays an axiomatic floor. On it he sets a hypothetical plank on which he raises a logical ladder which he proceeds to climb. There is risk, suspense and drama: any loose rung, any misstep fatal. At the proper confluence of space and time he steps off onto a higher platform with a broader panorama. The whole thing is fabrication—but so was creation.'

Katherine O'Brien

## 8.1 IN THE BEGINNING

The previous chapter represents the culmination of what computing *aficionados* might call the 'bottom-up' approach to the real numbers. That is to say, we began with the natural numbers ℕ and constructed, successively, the integers ℤ, the rational numbers ℚ and finally the real numbers ℝ. Now we examine the 'top-down' approach and adopt a more algebraic stance. The real numbers will be defined by an abstract set of axioms from which we shall deduce (as theorems), at first elementary, and later more sophisticated, properties of ℝ. The advantage of the axiomatic approach is that it does not depend on any preconceived ideas of what real numbers are. However, some of our first 'theorems' will appear trivial to our trained minds—the point to bear in mind is that they are consequences of even more basic assumptions— namely, our axioms. To make the following exposition more palatable, we shall present the defining axioms in three measured helpings. We shall also enlist the aid of our friends Alice, Tweedledee and Tweedledum... so, to work!

ℝ is a non-empty set upon which two closed binary operations $+$ and $\cdot$ are defined. In other words, given $x, y \in \mathbb{R}$, there exist in ℝ uniquely determined elements $x + y$ and $x \cdot y$. Since $+$ and $\cdot$ will turn out to be the familiar operations of addition and multiplication, we make no apologies for denoting our binary operations by these symbols. The set ℝ together with $+$ and $\cdot$ satisfy the following sets of axioms:

(1) The axioms of arithmetic.
(2) The axioms of order.

(3) The completeness axiom.

The precise nature of these axioms is contained in the following sections.

## 8.2 THE AXIOMS OF ARITHMETIC

A1 $a+(b+c)=(a+b)+c$
A2 $a+b=b+a$
A3 $\exists$ a unique element of $\mathbb{R}$, denoted by 0 such that $a+0=a$
A4 For any $a\in\mathbb{R}$, $\exists$ a unique element $x\in\mathbb{R}$ such that $a+x=0$
A5 $a\cdot(b\cdot c)=(a\cdot b)\cdot c$
A6 $a\cdot b=b\cdot a$
A7 $\exists$ a unique element of $\mathbb{R}$, denoted by 1, $1\neq 0$, such that $a\cdot 1=a$
A8 For any $a\in\mathbb{R}$, $a\neq 0$, $\exists$ a unique $y\in\mathbb{R}$ such that $a\cdot y=1$
A9 $a\cdot(b+c)=a\cdot b+a\cdot c$

*Notes*

(a) Each axiom is satisfied by any $a$, $b$ and $c$ in $\mathbb{R}$, except that $a\neq 0$ in A8.

(b) Any set satisfying A1–A9 is called a *field*. Those readers who have a knowledge of groups will observe that A1–A4 say that $\mathbb{R}$ is an Abelian group under $+$, and that A5–A8 say that $\mathbb{R}^*$ is an abelian group under $\cdot$. $\mathbb{R}^*$ is the set of real numbers excluding 0.

(c) The $x$ in A4 is called the *negative* of $a$ and will henceforth be denoted by $(-a)$.

(d) The $y$ in A8 is called the *reciprocal* of $a$ and will be denoted by $1/a$ or $a^{-1}$.

(e) A1 and A5 allow us to omit brackets in expressions such as $a+b+c$ or $a\cdot b\cdot c\cdot d$.

(f) Two particular numbers 0 and 1 and their roles are defined in A3 and A7. As yet we have no names for any other elements of $\mathbb{R}$.

(g) $0^{-1}$ is not defined—see A8.

(h) *Subtraction* is defined by $a-b=a+(-b)$.

(i) *Division* is defined by $a\div b=a/b$ or $a\cdot b^{-1}$, when $b\neq 0$.

The following sequence of elementary deductions from axioms A1–A9 shows us that our field $\mathbb{R}$ has elements which can be manipulated according to the familiar rules of school algebra. In essence, such algebra is none other than the axioms of arithmetic in disguise. However, with algebra we tend to accumulate new facts and learn various rules of thumb while at the same time forgetting that they are consequences of more elementary truths—this leads, unfortunately, to a seemingly unending list of ever more devious tricks. The structure of our mathematics is lost and inevitably many people develop a rather negative attitude towards mathematics.

Our first result seems facile! But note that it is not an axiom, since none of the axioms tell us anything about the behaviour of 0 relative to multiplication.

**Theorem 8.1** $x \cdot 0 = 0 \quad \forall x \in \mathbb{R}$.
*Proof* Now, $0 + 0 = 0$ by axiom A3. Multiplying both sides by $x$ gives $x \cdot (0 + 0) = x \cdot 0$. This is valid, since given two elements $a$ and $b$ of $\mathbb{R}$, their 'product' $a \cdot b$ is a uniquely determined element of $\mathbb{R}$. Applying axiom A9 yields $x \cdot 0 + x \cdot 0 = x \cdot 0$. Since $x \cdot 0$ is some element of $\mathbb{R}$, by axiom A4 it possesses a negative, $-(x \cdot 0)$. Hence,

$$(x \cdot 0 + x \cdot 0) + (-(x \cdot 0)) = x \cdot 0 + (-(x \cdot 0))$$

$$x \cdot 0 + (x \cdot 0 + (-(x \cdot 0))) = x \cdot 0 + (-(x \cdot 0)) \quad \text{using axiom A1}$$

$$x \cdot 0 + 0 = 0 \quad \text{using axiom A4}$$

$$x \cdot 0 = 0 \quad \text{using axiom A3} \quad \square$$

It is amazing that our axioms hid this well-known property of multiplication by zero. Notice that the first line of our proof confirms that $-0 = 0$; why? [**8.1**]

**Theorem 8.2** $x \cdot (-y) = -(x \cdot y) \quad \forall x, y \in \mathbb{R}$.
*Proof* $(x \cdot y) + x \cdot (-y) = x \cdot (y + (-y)) \quad \text{by A9}$

$$= x \cdot 0 \quad \text{by A4}$$

$$= 0 \quad \text{by Theorem 8.1}$$

But A4 tells us that the negative of $(x \cdot y)$ is unique and so we conclude that $-(x \cdot y) = x \cdot (-y)$. $\quad \square$

Ah! That was easier to prove, since we had Theorem 8.1 at our disposal. This could be the key to more rapid development, since we should be able to deduce more sophisticated results by using not only our axioms, but also previously established results. Perhaps, then, there is some logical order in which these elementary theorems of algebra should be deduced? If so, what is it, and how do we discover it? Almost by accident, as we do in the proof of the following.

**Theorem 8.3** $(-x) \cdot (-y) = x \cdot y \quad \forall x, y \in \mathbb{R}$.
*Proof* $(-x) \cdot (-y) = -((-x) \cdot y) \quad$ by replacing $x$ by $-x$ in Theorem 8.2

$$= -(y \cdot (-x)) \quad \text{by A6}$$

$$= -(-(y \cdot x)) \quad \text{by interchanging } x \text{ and } y \text{ in Theorem 8.2}$$

$$= -(-(x \cdot y)) \quad \text{by A6}$$

In order to complete the proof, all we now need to establish is the result that $-(-(a)) = a, \forall a \in \mathbb{R}$.

$$a + (-a) = 0 \quad \text{by A4}$$
$$(-a) + a = 0 \quad \text{by A2}$$
$$(-a) + (-(-a)) = 0 \quad \text{by A4}$$

But A4 tells us that the negative of $-a$ is unique and so we conclude that $-(-a) = a$, as desired. □

This property is the mysterious 'two minuses make a plus' or 'minus one times minus one equals plus one'. Look back at the intuitive models for $\mathbb{R}$ in Chapter 3 and find one that explains $(-1)\cdot(-1) = 1$! The favourite explanation is that it is defined that way. We have deduced it from the natural-looking axioms of arithmetic.

The proofs of the next two theorems are left as exercises. [8.2]

**Theorem 8.4** $\quad -(x^{-1}) = (-x)^{-1} \quad \forall x \in \mathbb{R}, \ x \neq 0$.
**Theorem 8.5** $\quad (x^{-1})^{-1} = x \quad \forall x \in \mathbb{R}, \ x \neq 0$.

We could continue our orgy of results, but the reader may rest assured that many of the arithmetical properties with which they are familiar can be deduced from axioms A1–A9. To conclude, we establish a slightly less elementary algebraic result.

**Example 8.1** $\quad (x - y) \cdot (x + y) = x^2 - y^2 \quad \forall x, y \in \mathbb{R}$.

*Proof*

| | |
|---|---|
| $(x - y)\cdot(x + y) = (x + (-y))\cdot(x + y)$ | see note (h) |
| $= (x + (-y))\cdot x + (x + (-y))\cdot y$ | by A9 |
| $= x\cdot(x + (-y)) + y\cdot(x + (-y))$ | by A6 |
| $= x\cdot x + x\cdot(-y) + y\cdot x + y\cdot(-y)$ | by A1 and A9 |
| $= x\cdot x + (-(x\cdot y)) + x\cdot y + (-(y\cdot y))$ | by A6 and Theorem 8.2 |
| $= x\cdot x + (x\cdot y + (-(x\cdot y))) + (-(y\cdot y))$ | by A1 and A2 |
| $= (x\cdot x + 0) + (-(y\cdot y))$ | by A1 and A4 |
| $= x\cdot x + (-(y\cdot y))$ | by A3 |
| $= x^2 - y^2$ | using note (h) |

and the usual shorthand $x^2$ for $x\cdot x$. □

As we can see, we are but a short step away from the algebraic rules for multiplication of brackets and factorisation in elementary algebra. Once there, the axioms A1–A9 will fade into the background and we find ourselves on the ground floor of a building called ALGEBRA. The foundations of this building are the axioms of arithmetic, but they are hidden from view.

## 8.3 THE AXIOMS OF ORDER

There is defined on $\mathbb{R}$ a relation $\leqslant$ satisfying

A10    $a \leqslant b \vee b \leqslant a$
A11    $a \leqslant b \wedge b \leqslant a \Rightarrow a = b$
A12    $a \leqslant b \wedge b \leqslant c \Rightarrow a \leqslant c$
A13    $a \leqslant b \Rightarrow a + c \leqslant b + c$
A14    $a \leqslant b \wedge 0 \leqslant c \Rightarrow a \cdot c \leqslant b \cdot c$

*Notes*

(a) Each axiom is satisfied by any $a, b, c \in \mathbb{R}$.

(b) A relation on a set is any two variable predicate of the form '$x$ is related (in some way) to $y$', where $x$ and $y$ are elements of the set. See Chapter 3 for a discussion of relations.

(c) Any set satisfying A1–A14 is called an *ordered field*.

(d) The relation $\leqslant$ is read as 'less than or equal to'.

(e) A13 and A14 tell us how $\leqslant$ behaves with respect to addition and multiplication.

(f) We can define on $\mathbb{R}$ a relation $<$, read as 'less than', by
$$a < b \Leftrightarrow a \leqslant b \wedge a \neq b$$

(g) The relations $\geqslant$ and $>$ ('greater than or equal to' and 'greater than', respectively) can easily be defined by
$$a \geqslant b \Leftrightarrow b \leqslant a$$
$$a > b \Leftrightarrow b < a$$

These five new axioms will now be used to deduce elementary properties of the order relation $\leqslant$. Just as the axioms of arithmetic defined the algebra of $=$ on $\mathbb{R}$, these axioms of order will define an algebra of $\leqslant$. The main use of $\leqslant$ will be to *order* the real numbers, but as yet we have names for relatively few real numbers; only 0, 1 and $-0$ (which is 0 again!) and $-1$. Show that $-1 \neq 0$. **[8.3]** Can $-1 = 1$? Reserve judgement on this until we have proved the following result.

**Theorem 8.6**    $0 < 1$.

*Proof*    By axiom A7, $0 \neq 1$ and so, in the light of note (f), it is sufficient to prove that $0 \leqslant 1$. We show that $1 \leqslant 0$ cannot hold and so, by A10, $0 \leqslant 1$ must be the case.

$\quad 1 \leqslant 0 \Rightarrow 1 + (-1) \leqslant 0 + (-1)$      by A13 and A4
$\quad\quad\quad\quad \Rightarrow 0 \leqslant -1$      by A2, A3 and A4
$\quad\quad\quad\quad \Rightarrow -1 \cdot 0 \leqslant (-1) \cdot (-1)$      using A14, with $a = 0$ and $b = c = -1$
$\quad\quad\quad\quad \Rightarrow 0 \leqslant 1$      by Theorems 8.1 and 8.3 and A7

But now we have $1 \leqslant 0$ (our original assumption) and $0 \leqslant 1$ (a deduction via the axioms). By axiom A11 we deduce that $0 = 1$, which violates axiom A7. □

This far from world-shattering result will shortly be used to deduce the familiar ordering of the integers. But the reader may now deduce that $-1 \neq 1$. [**8.4**]

We define $\mathbb{R}^+$ and $\mathbb{R}^-$, the sets of *positive* and *negative* real numbers, respectively, by

$$\mathbb{R}^+ = \{x : x \in \mathbb{R} \wedge x > 0\}$$

$$\mathbb{R}^- = \{x : x \in \mathbb{R} \wedge x < 0\}$$

By A10, $\mathbb{R} = \mathbb{R}^+ \cup \{0\} \cup \mathbb{R}^-$. By A11, $\mathbb{R}^+ \cap \mathbb{R}^- = \emptyset$. [**8.5**]

**Theorem 8.7** $x \in \mathbb{R}^+ \Leftrightarrow (-x) \in \mathbb{R}^-$.
*Proof*  $0 \leqslant x \Rightarrow 0 + (-x) \leqslant x + (-x)$     by A4, A13

$\Rightarrow (-x) \leqslant 0$     by A2, A3, A4

Conversely,

$(-x) \leqslant 0 \Rightarrow (-x) + x \leqslant 0 + x$     by A13

$\Rightarrow 0 \leqslant x$     by A2, A3, A4     □

When solving inequalities, we often wish to cancel common factors. Axiom A14 is of some use when we seek to cancel positive common factors. Our next theorem tells us that when cancelling negative common factors, we must *reverse the inequality sign*.

**Theorem 8.8** $x \leqslant y \wedge z \leqslant 0 \Rightarrow x \cdot z \geqslant y \cdot z$.
*Proof* Now

$$(-z) \geqslant 0 \quad \text{by Theorem 8.7}$$

So

$$x \cdot (-z) \leqslant y \cdot (-z) \quad \text{by A14}$$

So

$$-(x \cdot z) \leqslant -(y \cdot z) \quad \text{by Theorem 8.2}$$

We now add $(x \cdot z) + (y \cdot z)$ to both sides of the last inequality, invoke A13 and a bevy of the axioms of arithmetic to conclude that $(x \cdot z) \geqslant (y \cdot z)$. □

The proofs of the next three results are left as exercises. [**8.6**]

**Theorem 8.9** $0 \leqslant x \wedge 0 \leqslant y \Rightarrow 0 \leqslant x \cdot y$.
**Theorem 8.10** $0 \leqslant x \wedge y \leqslant 0 \Rightarrow x \cdot y \leqslant 0$.
**Theorem 8.11** $x \leqslant 0 \wedge y \leqslant 0 \Rightarrow 0 \leqslant x \cdot y$.

Similar results to the above can be proved with $<$ in place of $\leq$, but at this stage we have exposed the basic rules that can be applied when solving inequalities. We conclude by rather pedantically establishing a non-trivial inequality, using the axioms of order and the theorems above.

**Example 8.2** $a^2 + b^2 + c^2 \geq a \cdot b + b \cdot c + c \cdot a \quad \forall a, b, c \in \mathbb{R}$.

*Proof* First, note that as a consequence of Theorems 8.9 and 8.11, $x^2 \geq 0$ for any $x \in \mathbb{R}$. Also recall that in Example 8.1 we agreed to write $x^2$ as an abbreviation for $x \cdot x$. Now, for any $a, b, c \in \mathbb{R}$ we have that $(a-b)^2 \geq 0$, $(b-c)^2 \geq 0$ and $(c-a)^2 \geq 0$. Using A12 and A13 repeatedly, we deduce that

$$(a-b)^2 + (b-c)^2 + (c-a)^2 \geq 0$$

The left-hand side can be expanded, using axioms A1–A9, much as we did in Example 8.1, to obtain

$$2 \cdot ((a^2 + b^2 + c^2) - (a \cdot b + b \cdot c + c \cdot a)) \geq 0 \dagger$$

As a consequence of Theorems 8.9–8.11, we now deduce that

$$(a^2 + b^2 + c^2) - (a \cdot b + b \cdot c + c \cdot a) \geq 0$$

Freely using A1, A3, A4 and A13 yields the desired result. □

Before moving to the final axiom, which completes our axiomatic description of $\mathbb{R}$, we shall pause to describe certain subsets of $\mathbb{R}$. These we shall set in the context of the axioms of arithmetic and order. In particular, we shall gain names for many of the elements of $\mathbb{R}$ and then see the need for the final completeness axiom.

## 8.4 SUBSYSTEMS OF $\mathbb{R}$

Let $S \subseteq \mathbb{R}$ by any subset of $\mathbb{R}$ satisfying

(a) $1 \in S$
(b) $x \in S \Rightarrow x + 1 \in S$

For example, $S = \mathbb{R}^+$ satisfies (a) and (b). **[8.8]** The set of *natural numbers* $\mathbb{N}$ is defined to be the smallest subset of $\mathbb{R}$ satisfying (a) and (b). By smallest we mean that if $S$ is any subset of $\mathbb{R}$ satisfying (a) and (b), then $\mathbb{N} \subseteq S$. In fact, $\mathbb{N}$ could be defined as the grand intersection of all subsets of $\mathbb{R}$ satisfying

---

† 2 is used in the above proof as an abbreviation for $1 + 1$. As mentioned previously, we have not yet given names to many of our real numbers (only 0, 1 and $-1$). This will be remedied in Section 8.4. The properties of 2 that we used above were

(a) $2 \in \mathbb{R}^+$
(b) $x + x = 2 \cdot x, \forall x \in \mathbb{R}$

Property (b) is a simple application of A9, A6 and A7. Property (a) is left as an exercise. **[8.7]**

(a) and (b). Clearly, $1 \in \mathbb{N}$ and so is $1 + 1$, and $1 + 1 + 1$, etc.... Since $\mathbb{N}$ is the least such set satisfying (a) and (b), this process must lead to the enumeration of the elements of $\mathbb{N}$. In other words,

$$\mathbb{N} = \{1, 1 + 1, 1 + 1 + 1, \ldots\}$$

or, using commonly accepted labellings,

$$\mathbb{N} = \{1, 2, 3, \ldots\}$$

As can be seen in Chapter 5, this definition of $\mathbb{N}$ leads quickly to the first formulation of the powerful principle of mathematical induction. Finally, we can use Theorem 8.6 and axiom A13 to order the elements of $\mathbb{N}$ thus

$$1 \leqslant 2 \leqslant 3 \leqslant \ldots$$

the inequalities being, in fact, strict ones.

We can now define the set of *integers* $\mathbb{Z}$ by

$$\mathbb{Z} = \{0, \pm 1, \pm 2, \pm 3, \ldots\}$$

In other words, $\mathbb{Z}$ contains all the natural numbers, the negatives of each of the natural numbers and zero. Once again our axioms A1–A14 and their consequences allow us to order the integers

$$\ldots < -3 < -2 < -1 < 0 < 1 < 2 < \ldots$$

The final subsystem we define contains all possible products of integers and the reciprocals of non-zero integers. Formally the set of *rational numbers* $\mathbb{Q}$ is

$$\mathbb{Q} = \{m/n : m, n \in \mathbb{Z}, n \neq 0\}$$

Recall that $m/n = m \cdot n^{-1}$ and $n^{-1}$ is defined for $n \neq 0$. Notice that it is possible for different pairs of integers $m$ and $n$ to define the same rational number. See Chapter 3.

It is clear from their definitions that $\mathbb{N}$, $\mathbb{Z}$ and $\mathbb{Q}$ are all subsets of $\mathbb{R}$ and that $\mathbb{N} \subset \mathbb{Z} \subset \mathbb{Q} \subset \mathbb{R}$. We now show that $\mathbb{N}$, $\mathbb{Z}$ and $\mathbb{Q}$ are all closed under the binary operation of $+$.

## $\mathbb{N}$ is Closed under $+$

Pick any fixed $m \in \mathbb{N}$ and let $T$ be the set $\{n : n \in \mathbb{N} \wedge m + n \in \mathbb{N}\}$. That is, $T$ consists of those natural numbers whose sum with $m$ is still a natural number. $T$ satisfies condition (a) because $m \in \mathbb{N}$ implies $m + 1 \in \mathbb{N}$ by condition (b) for $\mathbb{N}$. $T$ also satisfies condition (b) because

$$\begin{aligned}
n \in T &\Rightarrow m + n \in \mathbb{N} &&\text{by the definition of } T \\
&\Rightarrow (m + n) + 1 \in \mathbb{N} &&\text{by condition (b) for } \mathbb{N} \\
&\Rightarrow m + (n + 1) \in \mathbb{N} &&\text{by axiom A1} \\
&\Rightarrow n + 1 \in T &&\text{by the definition of } T
\end{aligned}$$

Hence, $\mathbb{N} \subseteq T$, and since $T$ was defined as a subset of $\mathbb{N}$, $T = \mathbb{N}$. Essentially this is a proof by induction. □

## $\mathbb{Z}$ is Closed under $+$

By virtue of A3, adding 0 to an integer yields another integer. Addition of two positive integers (i.e. natural numbers) produces another integer (in fact, a natural number), by the result above. Now consider $n + (-m)$, where $n, m \in \mathbb{N}$.

If $n = m$, then $n + (-m) = 0$, an integer.

If $n > m$, then we show that there is a natural number $p$ such that $n = m + p$. The result is clearly true for $n = m + 1$ ($p$ merely equals 1). If for some particular $n > m$ there is a natural number $p$ satisfying $n = m + p$, then $n + 1 = (m + p) + 1 = m + (p + 1)$, using axiom A1. Since $p + 1 \in \mathbb{N}$ by condition (a) for $\mathbb{N}$, our result also holds for $n + 1$. Hence, our result holds for all $n > m$. This is an example of a proof by induction which does not start at $n = 1$. See Section 5.2.1. It follows immediately that for $n > m$, $n + (-m)$ is an integer. (In fact, $n + (-m)$ is the natural number $p$ whose existence we have just established.)

If $n < m$, then a similar careful argument yields that $n + (-m)$ is the negative of a natural number and so is an integer. [**8.9**] You will need to use the fact that $(-1) \cdot x = (-x)$, $\forall x \in \mathbb{R}$—a result worth adding to our theorems in Section 8.2. [**8.10**]

To complete the proof of the closure of $\mathbb{Z}$ under addition, we look at $(-n) + (-m)$, where $n, m \in \mathbb{N}$.

$$(-n) + (-m) = (-1) \cdot n + (-1) \cdot m$$
$$= (-1) \cdot (n + m)$$
$$= -(n + m)$$

which is the negative of a natural number and so lies in $\mathbb{Z}$. □

## $\mathbb{Q}$ is Closed under $+$

Given integers $m, n$ and $p$, where $n \neq 0$ and $p \neq 0$, and freely using the axioms of arithmetic,

$$m/n = m \cdot n^{-1} = m \cdot p \cdot p^{-1} \cdot n^{-1} = m \cdot p \cdot (n \cdot p)^{-1} = (m \cdot p)/(n \cdot p)$$

and so

$$m/n + M/N = (m \cdot N)/(n \cdot N) + (M \cdot n)/(N \cdot n) = (m \cdot N + M \cdot n)/(n \cdot N)$$

which is a rational number, since addition of integers is closed and provided that multiplication of integers is closed! Prove that multiplication defines a closed binary operation on $\mathbb{N}$, $\mathbb{Z}$ and $\mathbb{Q}$. [**8.11**]

We return now to the strict inclusions

$$\mathbb{N} \subset \mathbb{Z} \subset \mathbb{Q} \subset \mathbb{R}$$

and ask which of the axioms A1–A14 hold if $\mathbb{R}$ is replaced throughout by any subset of $\mathbb{R}$. In particular, we consider $\mathbb{N}$, $\mathbb{Z}$ and $\mathbb{Q}$. Our conclusions appear in Figure 8.1. At the end of Section 8.5 we shall look at other subsets of $\mathbb{R}$ in this light.

|   | A1 | A2 | A3 | A4 | A5 | A6 | A7 | A8 | A9 | A10 | A11 | A12 | A13 | A14 |
|---|----|----|----|----|----|----|----|----|----|-----|-----|-----|-----|-----|
| $\mathbb{N}$ | * | * |   |   | * | * | * |   | * | * | * | * | * |   |
| $\mathbb{Z}$ | * | * | * | * | * | * | * |   | * | * | * | * | * | * |
| $\mathbb{Q}$ | * | * | * | * | * | * | * | * | * | * | * | * | * |   |

* denotes that the axiom holds

**Fig. 8.1**

Axioms A1, A2, A5, A6, A9–A13 hold automatically for any subset of $\mathbb{R}$, since they are essentially formulae to be obeyed by all elements of $\mathbb{R}$. $\mathbb{N}$ and $\mathbb{Z}$ fail to be ordered fields but $\mathbb{Q}$ is an ordered field. Perhaps $\mathbb{Q} = \mathbb{R}$? No: in Chapter 4 we have seen that constructable lengths exist which are not rational. Our axiom system for $\mathbb{R}$ has not so far distinguished between $\mathbb{Q}$ and $\mathbb{R}$. Our final axiom will be the one that establishes that $\mathbb{Q} \neq \mathbb{R}$.

## 8.5 THE COMPLETENESS AXIOM

We have seen in previous chapters the deficiencies inherent in trying to describe the points on a number line with only rationals at our disposal. Our friends Alice, Tweedledee and Tweedledum have been watching our Axiomatic antics...

> TWEEDLEDUM: I can run faster than you!
> TWEEDLEDEE: No you can't!!
> ALICE: Come, come, no squabbling please; in any case I can run faster than anyone.
> TWEEDLEDUM: How can that be when you are so clearly shrinking?
> ALICE: Well, by the time you have run to where I am, I shall be somewhere else; and when you have reached 'somewhere else' I shall be gone again... ad infinitum... oh dear! I am shrinking awfully.
> TWEEDLEDUM and TWEEDLEDEE (*in unison*): But when you're caught, you're caught; so there!!

The Tweedles, far from having their brains addled by the preceding axiomatic anticlimaxes (they were only half listening, having never liked arithmetic at school), have almost resolved the paradox for us.

Suppose that Alice is given a start of $t_0$ metres. By the time Tweedledum (or Tweedledee) reaches Alice's starting point, she will have advanced another $t_1$ metres, say. As soon as Tweedledum reaches this further point, Alice is $t_2$ metres further on, and so on. See Figure 8.2. So, argues Alice, how can she

```
|————— t₀ —————|— t₁ —|-t₂-|
```

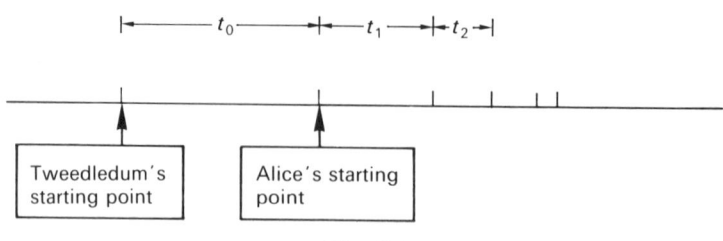

Fig. 8.2 *(diagram: number line showing Tweedledum's starting point and Alice's starting point, with intervals $t_0, t_1, t_2, \ldots$)*

be caught? As the brothers imply, there is a distance—$t$ metres, say—that one of them must run to catch Alice. $t$ is the *smallest distance* larger than all of the $t_0, t_0 + t_1, t_0 + t_1 + t_2, \ldots$. Notice that each of the partial sums $t_0, t_0 + t_1$, etc., could all be rational numbers; but as we shall see, $t$ need not be. Also note that for there to be a least number larger than all the partial sums, there must be some number which exceeds all of $t_0, t_0 + t_1$, etc. This motivates the following definitions.

A subset $S$ of $\mathbb{R}$ is *bounded above* if $\exists M \in \mathbb{R}$ such that $x \leqslant M$, $\forall x \in S$. $M$ is called an *upper bound* for $S$. For example, 0 is an upper bound for $\mathbb{R}^-$ (as indeed is 1 or 1.3). A subset $S$ of $\mathbb{R}$ is *bounded below* if $\exists m \in \mathbb{R}$ such that $x \geqslant m$, $\forall x \in S$. $m$ is called a *lower bound* for $S$. We can now state our final axiom—the completeness axiom.

A15 Every non-empty set of real numbers which is bounded above has a real smallest upper bound. Every non-empty set of real numbers which is bounded below has a real largest lower bound.

*Notes*

(a) Any set satisfying A1–A15 is called a *complete ordered field*.

(b) The smallest upper bound is called the *supremum* of the set and the largest lower bound is called the *infinum* of the set. They are unique (provided that they exist). [**8.12**]

(c) The second half of A15 is redundant. Can you see why? [**8.13**]

(d) The supremum or infinum of a set $S$ need not be elements of $S$.

Convince yourself that the sets and their suprema and infina in Figure 8.3 are correctly designated. [**8.14**]

Axiom A15 is precisely the one which guarantees the existence of irrational numbers. For example, consider the set

$$S = \{x : x \in \mathbb{Q} \wedge x^2 \leqslant 5\}$$

$S$ is bounded above (why?) and its supremum is none other than $\sqrt{5}$. Elements of $S$ are all rationals, but its supremum is not. Since the supremum is unique, $S$ is a set of rational numbers, bounded above, which has no rational supremum. Hence, A15 is *false* if we replace $\mathbb{R}$ by $\mathbb{Q}$; $\mathbb{Q}$ is not complete and also $\mathbb{Q} \neq \mathbb{R}$.

# AXIOMS FOR ℝ

| Set S | supremum of S | infinum of S |
|---|---|---|
| $\{1, 2, 3\}$ | 3 | 1 |
| $[1, 2)$ | 2 | 1 |
| $\mathbb{R}^+$ | does not exist | 0 |
| $\{1/(n^2+1): n\in\mathbb{Z}\}$ | 1 | 0 |
| $\{\sin(n\pi/2)/(n^2+1): n\in\mathbb{Z}\}$ | 1/2 | $-1/2$ |
| $\{x-\|x\|: x\in\mathbb{R}\}$ | 0 | does not exist |
| $\mathbb{Q}$ | does not exist | does not exist |
| $\{x: x\in\mathbb{R}, 0<x<1$ and the decimal expansion of $x$ contains no 9s$\}$ | 8/9 | 0 |

**Fig. 8.3**

**Exercise** Let $S = \{a + b\sqrt{5}: a, b\in\mathbb{Q}\}$. Show that $S$ is an ordered field. By showing that $\sqrt{7}\notin S$, deduce that $S$ is not complete. [**8.15**]

This completes our discussion of the axioms defining ℝ, but we finish this chapter by investigating some consequences of the completeness axiom. Suppose that one were tempted to say that ℝ had supremum $\infty$. After all, the symbol is liberally sprinkled about the literature. Is $\infty$ not a number bigger than all numbers, a sort of supremum of ℕ? If $\infty$ were a number, then presumably we would require that $\infty + \infty = \infty$. As a consequence of the axioms of arithmetic, we could deduce that

$$(\infty + \infty) - \infty = \infty - \infty = 0$$

So

$$\infty = 0$$

Help! Clearly this is nonsensical, so perhaps there are many different infinities $\infty, \infty'$, etc., with $\infty + \infty = \infty'$ perhaps? See Chapter 9. Our next result shows that the completeness axiom immediately implies that ℕ is not bounded above and so cannot possess a supremum.

**The Archimedean Property** ℕ is not bounded above.

*Proof* Suppose by way of contradiction that ℕ is bounded above by $B\in\mathbb{R}$. Hence, $n \leqslant B, \forall n\in\mathbb{N}$. By A15 we can assume that $B$ is the supremum of ℕ. Now $B-1$ cannot be an upper bound for ℕ and so $\exists m\in\mathbb{N}$ such that $m > B - 1$. Hence, $B < m + 1 \in \mathbb{N}$, which contradicts our choice of $B$. So ℕ cannot be bounded above. □

We are thus prevented from treating $\infty$ as a real number. Slightly more alarmingly, we cannot treat the infinitesimals $\triangle x$ and $\triangle y$ which arise in Newton's calculus as real numbers. An infinitesimal is taken to be an arbitrarily small non-zero quantity. Our final result shows that such quantities are not real numbers. This fact is one which renders imprecise the early development of calculus and resulted historically in the need for a rigorous basis to calculus—called analysis!

**Theorem 8.12** If $\beta\in\mathbb{R}$ satisfies $0 \leqslant \beta \leqslant x, \forall x\in\mathbb{R}^+$, then $\beta = 0$.

*Proof* We give a proof by contradiction, assuming that there is a $\beta$ satisfying $0 < \beta \leqslant x, \forall x \in \mathbb{R}^+$. Using the axioms of order, we soon deduce that

$$1/x \leqslant 1/\beta, \forall x \in \mathbb{R}^+$$

For each $n \in \mathbb{N}$ we let $x_n = 1/n \in \mathbb{R}^+$ and so

$$n = 1/x_n \leqslant 1/\beta, \forall n \in \mathbb{N}$$

In other words, $\mathbb{N}$ is bounded above, which contradicts the Archimedean property. Hence, $\beta = 0$. □

# 9 Some Infinite Surprises—in which some wild sets are tamed, and some nearly escape

'Je le vois, mais je ne le crois pas.'
                Georg Cantor, 1877

'Infinity is just so big that by comparison, bigness itself looks really titchy.'
        Douglas Adams, *The Hitch-hiker's Guide to the Galaxy*

## 9.1 ALICE AND COMPANY STUMBLE ON INFINITY

In this chapter we look at the properties, many of them paradoxical, of infinite sets. The subject is closely associated with the name of Georg Cantor (1845–1918). Unlike many, perhaps most, major theories in mathematics, Cantor's ideas on infinite sets owe little to foundations built over previous centuries by other mathematicians. He was the source of most of the ideas, and for this reason the subject is relatively easy to tie down to its origins. We shall be adding a few remarks to give some historical colour to our story, but by the end of the chapter you will probably agree that the subject is quite colourful enough anyway!

Infinite sets play an all-pervading role in mathematics, since most interesting sets are infinite—$\mathbb{N}$, $\mathbb{Q}$, $\mathbb{R}$, the set of all points in the plane, etc. We can investigate whether they are discrete (like $\mathbb{N}$) or dense (like $\mathbb{Q}$), are continuous (like $\mathbb{R}$) or have gaps (like $\mathbb{Q}$), are one- or two-dimensional, etc. Cantor was interested in a much more primitive and apparently trivial question: how many members does the set have? In particular, how are we to decide, of two given infinite sets, which is the bigger, or whether they are of the same size? Again we shall eavesdrop on a discussion of this problem between Alice and her friends, joining them while they are still considering finite sets.

TWEEDLEDEE: Last week I was helping my friend Bertrand Russell to write his masterpiece, *Principia Mathematica for the Unsuspecting*, which led me to a meditation on sets. What do you think of this problem?—

Decide which of the two sets $A$ and $B$ is the bigger.

$$A = \{\text{Alice, Gryphon, Mock Turtle}\}$$

and

$$B = \{\text{Dormouse, Mad Hatter, March Hare, White Rabbit}\}$$

TWEEDLEDUM: $A$, of course; Gryphon could eat everything in $B$ and not get much bigger.
ALICE: I don't think Dee means that sort of big, Dum. He's not really concerned about how big the Gryphon is and how small the Dormouse is.
TWEEDLEDEE: No, of course not; Dum's just being difficult as usual. I mean, which set has got the biggest number of members?
TWEEDLEDUM: I wish you'd said that! $B$ has 4, $A$ has 3, 4 is bigger than 3, so $B$ is the bigger set. How very exciting!
TWEEDLEDEE: There's no need for your sarcasm. If you'd been better educated, you'd know that decent philosophy starts from simple things but soon ties you in knots quite effectively.
ALICE: I'm just a bit worried about the Dormouse, Dee. He's not very good at counting but I'm sure he'd have no difficulty telling which was the bigger set. How do you think he does it?
TWEEDLEDEE: Easy! Haven't you heard him humming dance tunes whenever he does this sort of problem? He dances with you, Mad Hatter with Gryphon and March Hare with Mock Turtle. White Rabbit is left without a partner, and he's in set $B$, so $B$ must be the bigger set.
ALICE: That's pretty clever of him! He wouldn't even need to know the names of the numbers, let alone know how to count.
TWEEDLEDEE: It's so nice to meet perceptive people, Alice. What do you think of this method? If $A$ is $\{a, e, i, o, u\}$ and $B$ is $\{a, b, c, d, e, i, o, u, z\}$ $B$ must be the bigger set because everything in $A$ is in $B$ and there are some bits of $B$ not in $A$.
TWEEDLEDUM: I'm getting the idea now. That's a good method, too, since it's very primitive—nothing sophisticated like counting in it.
ALICE: We've got three methods now. I wonder whether they all come down to the same thing really, or whether you can't always apply all the methods, or whether sometimes they give different answers, or....

To cut a long story short, let us start to look at some of Alice's speculations. Clearly, if we are just presented with a couple of sets 'at random', the chances are that neither will be a subset of the other, so this effectively condemns Dee's final suggestion as a general method, even for finite sets. The first method, Dum's simple-minded 'count and compare', is fine for finite sets, but by definition we can never 'count' all the elements of an infinite set.

This leaves the Dormouse's dancing partners method, and it was one of Cantor's early achievements to realise that this method could be used to give a sensible answer to the question, 'How many members?' even for infinite sets. In slightly more formal language than Dum's, the principle is that if each element of *A* can be paired with an element of *B*, no element of either set partnering more than one of the other, and no element of either set left over, then *A* and *B* can legitimately be called *equinumerous*. It is, of course, just the principle of the one-to-one correspondence or bijection which we discussed in Chapter 6.

So our final definition is:

*A* and *B* are equinumerous if there exists a bijection between them.

This looks fairly innocuous, but if we accept it, we quickly commit ourselves to some pretty strange consequences, as the inhabitants of Wonderland soon discovered. Fortunately, as we shall see, they did manage to come to terms with the strangeness.

TWEEDLEDUM: I agree that your bijections are ingenious, but have you really thought it through? What about this? [He shows the others the scheme below.]

$$\begin{array}{cccccccc} 1 & 2 & 3 & 4 & 5 & 6 & 7 & \ldots \\ \updownarrow & \updownarrow & \updownarrow & \updownarrow & \updownarrow & \updownarrow & \updownarrow & \ldots \\ 2 & 4 & 6 & 8 & 10 & 12 & 14 & \ldots \end{array}$$

You really want to say on the basis of this that there are just as many *even* natural numbers as there are natural numbers? I can see that my method doesn't work in general, but it does in this case. The even numbers are only half of $\mathbb{N}$, but you want me to believe the two sets are the same size! You are perverse, to say the least.

ALICE: Yes, I can see you have a point, but it seems to depend on grasping 'half of infinity'. I'm not sure that I could make that understandable.

TWEEDLEDEE: Yes, I agree that Dum has a point too. There is such an enormous pressure from intuition to say that the evens are clearly a proper subset of the naturals, so to claim they are equinumerous must be crazy. But I'm going to ask you to keep an open mind for a couple of minutes because I'm sure I have the antidote to that feeling.

TWEEDLEDUM: Carry on. Any clarification would be welcome. I think I can feel an attack of confusion coming on.

TWEEDLEDEE: I think you're still uneasy because you can't get away from the identity of the *elements* of the sets (Dum's original mistake). Remember that we're trying to capture the idea of the *size* of a set in the sense of *how many elements* it has, nothing to do with what the elements are. Now look again at the set $\{2, 4, 6, 8, \ldots\}$. We could just as well call it $\{E_1, E_2, E_3, E_4, \ldots\}$—I'm thinking of {first even number, second even number,...}, and that clearly doesn't change the size of the set.

ALICE:  
TWEEDLEDUM: } Agreed.

TWEEDLEDEE: Good, now will you just accept one more thing like that? The size of the set also has nothing to do with what we *call* the set, so we can forget about the nature of the elements, their *Even*-ness, and just concentrate on their name-tags, so we arrive at $\{1, 2, 3, 4, \ldots\}$, a set which is clearly equinumerous with $\mathbb{N}$ on any reasonable definition.

ALICE: Thank you, Dee. I enjoyed that argument. Incidentally, one of my other worries has disappeared too. If you'd just shown me that there was a bijection between $\mathbb{N}$ and the evens, I don't think I would have minded at all. I just got uneasy when you insisted on calling this 'equinumeracy'.

TWEEDLEDUM: I quite agree, Alice. Just a language problem again. Didn't Shakespeare say, 'The fault, dear Alice, lies not in our ideas but in our words'?

ALICE: Well, not exactly that, Dum. Don't worry, it's only your words at fault. The idea's exactly what I had in mind.

The Wonderland trio seem to have arrived at an agreement that there is nothing wrong with the reasoning which led to the establishment of a bijection between $\mathbb{N}$ and $E$ (where $E$ denotes the set of even naturals), but that to signal this by the words '$\mathbb{N}$ and $E$ are equinumerous' is out of step with usual usage of this phrase. Logically, of course, it doesn't matter: if you don't like this usage, all you have to do is mentally translate any sentence of the form '$A$ and $B$ are equinumerous' into 'there is a bijection between $A$ and $B$', and regard the whole theory as being about properties of bijections rather than about the size of sets. You, the reader, may choose. We shall continue to use the more colourful language.

Historically the issue was perceived as important, and the fact that infinite sets turned out to be equinumerous with some of their proper subsets was a cause for much suspicion in the early years of the theory. With hindsight, perhaps this was to be expected: we are trying to extend the scope of a concept, 'size of a set' from finite sets, which we feel we know all about, to infinite sets, which behave similarly to but not identically with finite sets, so the analogy cannot be perfect. What we need to keep hold of is the thought that when the context of a question like 'How many members does $\{a, e, i, o, u\}$ have?' is widened to ask questions like 'How many members does $\mathbb{N}$ have?' we must expect some answers which look paradoxical when viewed from the original narrower context. (Remember the similar discussion we had about the meaning of $\Rightarrow$ in Chapter 1.)

Let us hold on to this method of capturing 'the size of a set' and see where it leads us. We shall write $A \leftrightarrow B$ to mean that there is a bijection from $A$ to $B$, which is our extension from finite sets to all sets of the notion of two sets having the same number of elements. You should convince yourself that the two ideas do actually coincide when the sets are both finite. [**9.1**] We have already discussed at some length one way in which our use of language is perverse, so we should check that it is not totally perverse. The existence of a bijection from $A$ to $B$ is meant to indicate some degree of similarity between the sets—a similarity in 'size' in this case. We list below three properties we would expect of anything purporting to be 'similarity'.

(1) 'Being identical with' is clearly a stronger notion that 'being similar to', so we would expect any set to be similar to itself!

(2) Similarity is usually understood to be a mutual property. If $A$ were judged similar to $B$, it would be perverse to judge $B$ not similar to $A$.

(3) Similarity should be consistent in the sense that if $A$ is similar to $B$, and $B$ to $C$, then $A$ should be similar to $C$.

These expectations, translated into properties expected of $\rightsquigarrow$, are:

(1) $A \rightsquigarrow A$.
(2) $A \rightsquigarrow B \Rightarrow B \rightsquigarrow A$.
(3) $(A \rightsquigarrow B \wedge B \rightsquigarrow C) \Rightarrow A \rightsquigarrow C$.

To illustrate what is involved we prove property (3).

Let $f$ and $g$ be bijections from $A$ to $B$ and $B$ to $C$, respectively. We have to show there is a bijection from $A$ to $C$. A fairly obvious candidate is the map $m$ from $A$ to $C$ defined by $m(a) = g(f(a))$ for each $a$ in $A$. To show that this really is a bijection from $A$ to $C$ three things need to be established.

(1) Each $a \in A$ is mapped to a unique member of $C$.
(2) Each $c \in C$ is the image of something in $A$.
(3) No member of $C$ can be the image of more than one member of $A$.

Condition (1) is obvious from the specification of $m$. $c$ must be $g(b)$ for some $b$ in $B$, since $g$ is a bijection, and $b$ must be $f(a)$ for some $a$ in $A$, since $f$ is a bijection. Hence, $c = g(f(a)) = m(a)$, so condition (2) is established.

If $c$ were the image under $m$ of $a_1$ and $a_2$, we would have $g(f(a_1)) = g(f(a_2))$, so $f(a_1) = f(a_2)$, since $g$ is a bijection. This, in turn, implies $a_1 = a_2$, since $f$ is a bijection, so condition (3) is established. □

In Chapter 3 we had the logical problem of saying what we meant by 'integer' when our official knowledge only covered $\mathbb{N}$. We eventually arrived at the idea of defining, for example, $-5$ as a set of ordered pairs of natural numbers, and the sets of ordered pairs we wanted to use were equivalence classes of the equivalence relation $\approx$.

We now have a similar situation: our aim was to say what we mean by the size (or cardinal number, to give it its accepted title) of an infinite set, when bijections are available in our official knowledge. We have just shown that $\rightsquigarrow$ is an equivalence relation between sets, so we can use the same device. For example, the cardinal number usually called 3 is the equivalence class containing $\{a,b,c\}$, or the set of all sets $A$ having the property that $A \rightsquigarrow \{a,b,c\}$. This is essentially the definition used by Russell and Whitehead in their logistic treatment of natural numbers in 1910. It is the method used by Cantor to define infinite numbers too: the cardinal number of $\mathbb{N}$ is the set of all sets in one-to-one correspondence with $\mathbb{N}$.

## 9.2 THE SMALLEST INFINITY

We shall use the terms, 'denumerable' to mean equinumerous with $\mathbb{N}$, and 'countable' to mean either finite or denumerable. Our notation for cardinal numbers will be $\bar{\bar{A}}$ for the size of set $A$, so $\bar{\bar{A}} = \bar{\bar{B}}$ will mean $A \leftrightarrow B$. The double bar notation is Cantor's original one and there is a good reason for it. In subjects such as analysis the basic set is $\mathbb{R}$, and the fact that this set and its related operations make what is called a complete ordered field is of vital importance. On the other hand, in the theory of abstract ordered sets (the theory of ordinal numbers) the nature of the elements is of no importance; only their *order* is significant. Going further, the theory of cardinal numbers strips away even this amount of structure and retains only the *number* of elements in the sets. It concerns itself with neither the nature nor the ordering of the elements, and Cantor's two bars refer to this double act of abstraction.

In his first paper on the subject (in 1874) he proved several basic properties of countable sets. Their proofs are simple, though often ingenious, and they are still the starting points of modern accounts of the theory. We now look at some of these results, their proofs and their rather striking colloquial interpretations.

> **Theorem 9.1**  Any subset of a countable set is countable.

Clearly, any subset of a finite set is countable and any finite subset of a countable set is countable. The remaining claim of the theorem is that any infinite subset of a denumerable set is denumerable. This is where the interest of the theorem lies, because it claims that $\mathbb{N}$ is the 'smallest' infinite set.

To prove it, let $A$ be our denumerable set and $B$ any infinite subset of $A$. By definition there is a bijection $f$ from $\mathbb{N}$ to $A$, so the members of $A$ are $f(1), f(2), f(3), \ldots$. $B$ consists of some of these, so if we consider any $b \in B$, it will occur somewhere in the list: if $n$ other elements of $B$ precede it, we assign the natural number $n + 1$ to $b$. In this way each element of $B$ is assigned a unique element of $\mathbb{N}$, and vice versa. That is, we have a bijection between $B$ and $\mathbb{N}$, which means that $B$ is denumerable.  □

Now, in what sense does this mean that $\mathbb{N}$ is the smallest infinite set? It is just that if we start with any denumerable set and try to make it smaller by throwing out some of its members, either we must not throw out enough, and still end up with a denumerable set, or we overdo it and reduce it to a finite set. There is no way of getting in between these extremes!

Having raised the issue of comparison of sizes, it is natural to raise the question of what it could mean to say that one set is smaller than another. In other words, what is a sensible definition of $\bar{\bar{A}} < \bar{\bar{B}}$? If $A$ and $B$ are both finite, there is no problem: $\bar{\bar{A}} < \bar{\bar{B}}$ can be defined as the existence of a bijection from $A$ to a proper subset of $B$. If we tried this for infinite sets, it would have the unfortunate consequence that $\bar{\bar{E}} < \bar{\bar{N}}$, $\bar{\bar{E}} = \bar{\bar{N}}$ and $\bar{\bar{N}} < \bar{\bar{E}}$ would *all* be true! Convince yourself that this is so. [**9.2**] We can avoid this by adding a clause to our first attempt, to arrive at: $\bar{\bar{A}} < \bar{\bar{B}}$ means that there is a bijection from $A$ to a subset of $B$ and there is no bijection from $A$ to the whole of $B$.

Then we have ensured (by definition) that $\bar{\bar{A}} = \bar{\bar{B}}$ and $\bar{\bar{A}} < \bar{\bar{B}}$ can't both be true. What do you think of the possibility of $\bar{\bar{A}} < \bar{\bar{B}}$ and $\bar{\bar{B}} < \bar{\bar{A}}$ both being true? [**9.3**] More food for thought is whether, for arbitrary sets $A$ and $B$, at least one of $\bar{\bar{A}} = \bar{\bar{B}}$, $\bar{\bar{A}} < \bar{\bar{B}}$, $\bar{\bar{B}} < \bar{\bar{A}}$ must be true. [**9.4**] Part of what makes infinity interesting is that the answers to both these latter questions are very far from obvious.

Our next theorem is even more striking.

> **Theorem 9.2** If $A$ is infinite and $B$ is countable then $\bar{\bar{A}} = \overline{\overline{A \cup B}}$.

This can be interpreted as saying that if we start with an infinite set and throw in extra elements whose totality is no more than denumerably infinite, we don't change the size of the set!

Before proving this, notice that the statement of the theorem says nothing about the relationship between $A$ and $B$. In particular, they may overlap or even, to take an extreme case, be the same set. In this latter case the theorem reduces to $\bar{\bar{A}} = \overline{\overline{A \cup A}}$, and since $A = A \cup A$, this could hardly be called an exciting revelation. The other extreme of $A$ and $B$ being totally disjoint sets is obviously more difficult to prove, and the intermediate cases where there is partial overlap between $A$ and $B$ may be even harder, because it would seem to depend on the precise degree of overlap that existed (the size of $A \cap B$).

Many proofs in mathematics texts begin with the magic incantation 'Without loss of generality we may assume that ...'. In the case of our theorem you may well find 'without loss of generality we may assume that $A$ and $B$ are disjoint', and it is worth pausing to discuss this common abuse of language. It is an abuse because, clearly, it does not mean what it says: to claim that $\bar{\bar{A}} = \overline{\overline{A \cup B}}$, for *any* infinite $A$ and *any* countable $B$ is to make a stronger claim than $\bar{\bar{A}} = \overline{\overline{A \cup B}}$ only for infinite $A$'s and countable $B$'s with a special property—namely, that $A \cap B = \varnothing$. So, generality certainly has been lost in proving the weaker claim only. What is usually meant is that, having proved the weaker claim, the stronger more general one can be deduced easily. This is the case here, because $A \cup B$ can always be written as $A \cup (B \setminus A)$ and $A, B \setminus A$ are certainly disjoint by definition of $\setminus$, and $B \setminus A$ is still countable by the previous theorem.

With that little (but important) digression out of the way, we can get on with the proof. All we need to do is find some bijection from $A$ to $A \cup B$, where $A$ is infinite, $B$ countable and $A \cap B = \varnothing$. It is convenient to consider two separate cases.

*Case 1:* $B$ is denumerable
In this case $B$ can be considered as an infinite sequence $\{b_1, b_2, b_3, \ldots\}$. We then extract any denumerable subset $A^*$ from $A$ and call its members $\{a_1, a_2, a_3, \ldots\}$.

Now define the function $f: A \to A \cup B$ by

$$f(x) = \begin{cases} a_n & \text{if } x = a_{2n} \\ b_n & \text{if } x = a_{2n-1} \\ x & \text{if } x \in A \setminus A^* \end{cases}$$

In other words, $a_2, a_4, a_6$ are mapped to $a_1, a_2, a_3, \ldots$, respectively and $a_1, a_3, a_5$ are mapped to $b_1, b_2, b_3, \ldots$, respectively. This defines a bijection from $A^*$ to $A^* \cup B$, and the third line of the definition extends this to those members of $A$ not in $A^*$ (which stay fixed), so we see that $f$ is a bijection from $A$ to $A \cup B$ as required. This is illustrated in Figure 9.1.

*Case 2:* $B$ is finite
Define $A^*$ as in case 1 and suppose that $B$ is $\{b_1, b_2, \ldots, b_k\}$. We define $g: A \to A \cup B$ as follows:

$$g(x) = \begin{cases} a_n & \text{if } x = a_{2n} \text{ and } n \leq k \\ b_n & \text{if } x = a_{2n-1} \text{ and } n \leq k \\ a_{k+l} & \text{if } x = a_{2k+l} \text{ and } l \geq 1 \\ x & \text{if } x \in A \setminus A^* \end{cases}$$

It is probably easier to see that this is a bijection if we write out in detail what it does to the elements of $A$:

$a_2, a_4, a_6, \ldots, a_{2k}$ are mapped to $a_1, a_2, a_3, \ldots, a_k$, respectively

$a_1, a_3, a_5, \ldots, a_{2k-1}$ are mapped to $b_1, b_2, b_3, \ldots, b_k$, respectively

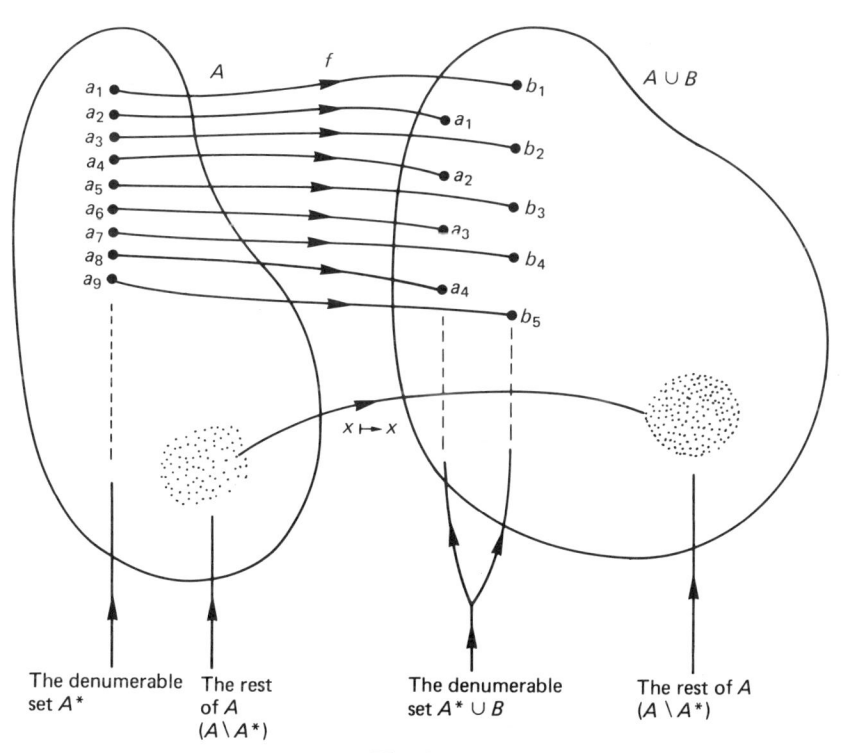

**Fig. 9.1**

$a_{2k+1}, a_{2k+2}, a_{2k+3}, \ldots$ are mapped to $a_{k+1}, a_{k+2}, a_{k+3}, \ldots$, respectively and everything in $A \setminus A^*$ stays fixed, as in Figure 9.2. □

An easy corollary of this is that the union of any finite number of countable sets is countable. [**9.5**]

The simple picture in Figure 9.3 yields a result with many ramifications. It represents denumerably many rows of dots, each row containing denumerably many dots. The resulting infinite rectangular array can be thought of either as a denumerable union of disjoint denumerable sets (each row containing the elements of one of the sets) or as the Cartesian product of two denumerable sets (if the sets are $A$ and $B$, with $A = \{a_1, a_2, a_3, \ldots\}$ and $B = \{b_1, b_2, b_3, \ldots\}$, then the element $(a_i, b_j)$ of $A \times B$ is represented by the dot in the $i$th row and the $j$th column). The arrows indicate how a path which visits each dot once and once only may be traced through the array. If we list the dots in the order in which they are visited, the whole array is thereby expressed as a single denumerable set.

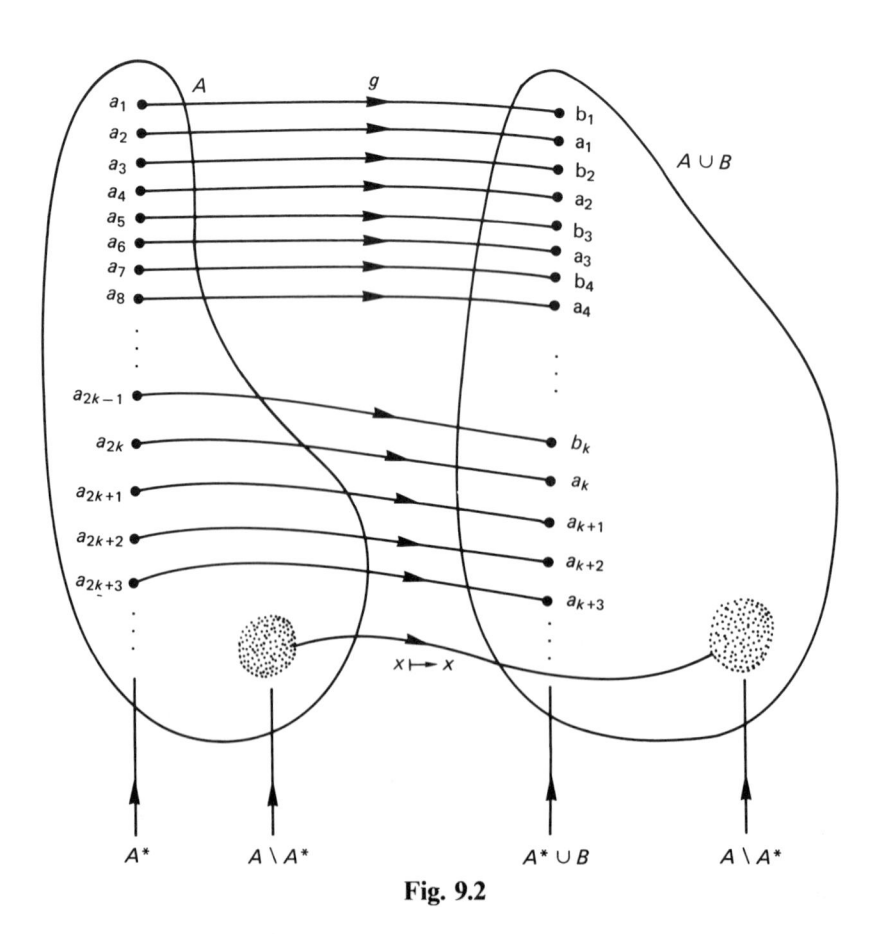

**Fig. 9.2**

Fig. 9.3

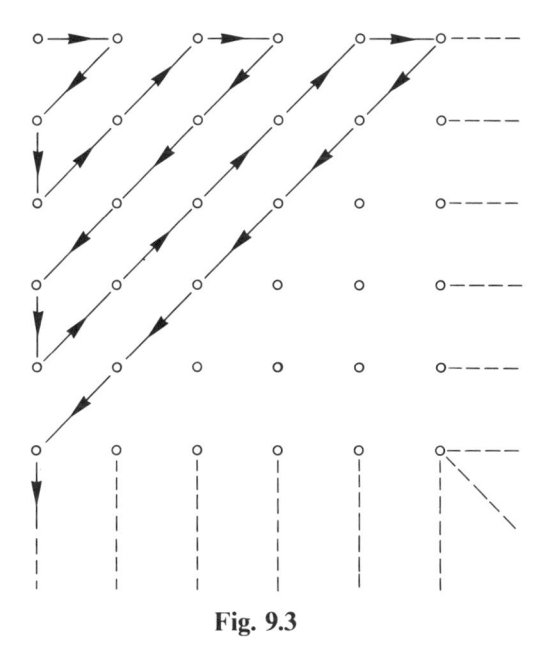

This simple observation directly yields the following pair of equivalent theorems.

---

**Theorem 9.3(a)**  If $A$ and $B$ are denumerable, so is $A \times B$.
**Theorem 9.3(b)**  If $\{A_i : i = 1, 2, 3, \ldots\}$ is a denumerable set of denumerable sets which are pairwise disjoint, then $\cup_{i \in \mathbb{N}} A_i$ is denumerable.

---

(Pairwise disjoint just means that no distinct pair of $A_i$'s have any element in common.)

These theorems are easily generalised by replacing all the 'denumerables' by 'countables'. In the case of Theorem 9.3(a) our array would then only have a finite number of rows or columns, or both, and for Theorem 9.3(b) some of the rows may only have a finite length and/or there may only be finitely many of them. In any of these cases the path through the array can be adjusted easily. [**9.6**]

Another generalisation is to drop the disjointness condition in Theorem 9.3(b). In terms of the array, this would mean that several dots could represent the same member of $\cup A_i$. To deal with this we could maintain the same route through the array but simply omit any dot which represented a member of $\cup A_i$ previously visited.

An alternative method of dealing with the generalisation from denumerable to countable is to say that this corresponds to going from the full array to a subset of it, and then appeal to Theorem 9.1. [**9.7**]

Finally, Theorem 9.3(a) can be further generalised by induction to the Cartesian product of any finite number of countable sets. [**9.8**]

The most famous consequence of these results is that $\mathbb{Q}$ is a denumerable set. The proof is not difficult: think about how $\mathbb{Q}$ is defined, then you can use Theorem 9.3(a) or Theorem 9.3(b). [**9.9**]

We have already spent considerable time in previous chapters on such things as the density of $\mathbb{Q}$, so what is our attitude to the denumerability of $\mathbb{Q}$? Should we be surprised and excited by the fact that $\mathbb{N}$, such an apparently minute subset of $\mathbb{Q}$, has the same size as $\mathbb{Q}$, or should we begin to lose faith (or at least interest) in Cantor's method on the grounds that it can't even distinguish between $\mathbb{N}$ and $\mathbb{Q}$? Your answer, of course, depends on how much you know about the theory and how much experience you have had using it. We reserve judgement at this stage, and merely point out that this and similar dilemmas gave Cantor's brain-child a very rough infancy at the hands of some other mathematicians.

## 9.3 A BIGGER INFINITY

We could go on in this way and produce a large number of apparently very big sets, all of which would turn out to be denumerable. It would be possible to produce such an impressive list that you could be forgiven for guessing that every infinite set is denumerable, in which case the whole theory of sizes of sets would be dull, to say the least. Such a course will be more interesting after seeing that non-countable sets exist, and in view of the fuss we have made over $\mathbb{R}$, you may have guessed that $\mathbb{R}$ is indeed the significant example.

> **Theorem 9.4** $\mathbb{R}$ is not denumerable.

How do we show this? The basis of our method is to take an arbitrary denumerable subset of $\mathbb{R}$ and show that it cannot contain every real number. Hence, $\mathbb{R}$ itself cannot be denumerable. The proof is constructive in the sense that from our denumerable subset of $\mathbb{R}$ we actually give a procedure for finding a real number not in this set.

We have mentioned several ways of representing real numbers in Chapter 4, and we choose to use nests for this application.

*Details of the Proof* First enumerate the chosen denumerable subset $S$ of $\mathbb{R}$ by $r_1, r_2, r_3, \ldots$. Then for each $r_i$ choose a nest $\{I_{in}\}$ to represent it, where the sequence of closed intervals which make up $\{I_{in}\}$ is $I_{i1}, I_{i2}, I_{i3}, \ldots$. To make our proof work smoothly, we make the $i$th interval, $I_{ii}$, of length less than $3^{-i}$, for each $i$.

We now construct another nest $\{J_n\}$ which is not equivalent to any of

# SOME INFINITE SURPRISES

the $\{I_{in}\}$. This means that the real number represented by $\{J_n\}$ will be different from all the $r_i$, as required.

Step 1: choose $J_1$ disjoint from $I_{11}$, and of length 1
Step 2: choose $J_2 \subset J_1$ with $J_2 \cap I_{22} = \varnothing$, with length $3^{-1}$
Step 3: choose $J_3 \subset J_2$ with $J_3 \cap I_{33} = \varnothing$, with length $3^{-2}$
...
..., and so on

You should now make sure that each step can actually be carried out—that is, check that, given $J_n$, the interval $J_{n+1}$ with the properties specified in step $n+1$ does exist. [**9.10**] Having verified this, we see that, for each $n$, $J_{n+1} \subset J_n$ and that by construction their lengths approach zero because the length of $J_n$ is $3^{-n}$. So the $J_n$'s do form a nest, and, hence, represent a real number $r$. It remains to show that $r$ is a different real number from all the $r_i$. To do this, note that our construction ensures that $J_i \cap I_{ii} = \varnothing$. Hence, by our criterion of Chapter 4, the nests $\{J_n\}$ and $\{I_{in}\}$ are not equivalent. In other words, $r$ and $r_i$ are different real numbers. □

## 9.4 A FEW REMARKS ON WHAT HAS BEEN DONE SO FAR

### 9.4.1 What *Does* 'Denumerable' Mean?

Technically we know the answer to this question: $D$ is a denumerable set if a bijection exists between $D$ and $\mathbb{N}$. In practice we have often taken it to mean that the elements of $D$ can be written out as an infinite list, and there seems to be some sleight of hand going on here, because surely, if we are allowing ourselves the luxury of infinitely long lists, then *any* set can be listed! To understand why this is not so, it is important to realise how we are restricting the meaning of 'list'.

For example, consider these two 'infinitely long listings' of the integers:

(a) $1, 2, 3, 4, \ldots, 0, -1, -2, -3, -4, \ldots$

and

(b) $0, -1, 1, -2, 2, -3, 3, -4, 4, -5, \ldots$

List (a) is certainly a well-defined ordering of $\mathbb{Z}$ in the sense that, given any two integers $x$ and $y$, it is possible to say whether $x$ comes before or after $y$. To be precise, $x$ precedes $y$ in exactly these cases: $x$ and $y$ are both natural numbers and $x < y$, or $x$ and $y$ are both outside $\mathbb{N}$ and $x > y$, or $x$ is in $\mathbb{N}$ and $y$ is not.

Similarly, in (b) $x$ precedes $y$ if and only if $|x| < |y|$, or $|x| = |y|$ and $x < 0$.

The difference is that list (b) represents a bijection from $\mathbb{N}$ to $\mathbb{Z}$ but (a) does not. The convention employed implicitly in the preceding parts of this chapter, which we now make explicit, is that in representing a bijection as a list 1 is mapped to the first member of the list, 2 to the second, and so on. In (b) it is clear that, given any integer $x$, $x$ will be the $n$th member of the list for some natural number $n$. (It is even possible to give a formula for $n$ in terms of $x$ in this case [**9.11**]). But in (a) this cannot be done. Which natural number does $-3$ correspond to, for example?

To show that $\mathbb{Z}$ is denumerable, all we have to do is produce one ordering which corresponds to a bijection with $\mathbb{N}$, and this is achieved in (b). Both attempts (a) and (b) were guided by the 'natural' order of $\mathbb{Z}$—namely $\ldots -3, -2, -1, 0, 1, 2, 3, \ldots$—but when we come to $\mathbb{Q}$ there is no natural order, in the sense that, given a rational number $q$, there is no 'next biggest' rational, $r$. Why not? [**9.12**] However, $\mathbb{Q}$ can be listed to give a bijection as follows:

$$\frac{0}{1}, \frac{1}{1}, \frac{-1}{1}, \frac{2}{1}, \frac{-2}{1}, \frac{1}{2}, \frac{-1}{2}, \frac{3}{1}, \frac{-3}{1}, \frac{1}{3}, \frac{-1}{3}, \frac{4}{1}, \frac{-4}{1}, \frac{3}{2}, \frac{-3}{2}, \frac{2}{3}, \frac{-2}{3}, \frac{1}{4}, \ldots$$

How was this ordering produced and how can you convince yourself that it is a bijection? [**9.13**]

You should now be in a better position to appreciate the full force of Theorem 9.4. It means that however we try to order $\mathbb{R}$, we can *never* achieve a bijection with $\mathbb{N}$.

### 9.4.2 Is the Method of Proof that $\bar{\mathbb{R}} > \bar{\mathbb{N}}$ Just a One-off Trick?

Clearly no! If it were, we wouldn't bother to raise the question, but what is interesting is that the method has such a wide range of application. As well as showing that certain sets (like $\mathbb{R}$) are bigger than $\mathbb{N}$, it can be adapted to show that some sets are not finite, that some are bigger than $\mathbb{R}$, and it even has a purely finite version. Here are some of these applications in action.

*How Many Primes are There?* Well, the primes are a subset of $\mathbb{N}$, so they are certainly countable, and we have to decide whether there are only a finite number or not. If we think of the primes as dots at the appropriate positions on the number line, they certainly thin out as we explore the number line further to the right. This is perhaps not too surprising, since, if we pick a big number, there are lots of smaller ones which may be factors of it, but if we pick a smaller one, there is a fair chance that its relatively smaller number of predecessors will not include a factor of it (except 1). This expectation is confirmed by the numerical evidence displayed in Figure 9.4. The table gives an 'average percentage density' of primes at various parts of the number line. The fourth entry, for example, indicates that, if we take a sample of a

# SOME INFINITE SURPRISES

| n: | $10^3$ | $10^6$ | $10^9$ | $10^{12}$ | $10^{20}$ | $10^{30}$ | $10^{60}$ |
|---|---|---|---|---|---|---|---|
| Average % density of primes near n | 14 | 7 | 5 | 3.5 | 2 | 1.4 | 0.7 |

**Fig. 9.4**

hundred numbers in the neighbourhood of a million million, then on average we would find that three or four of them were prime.

In fact, it is easy to show that there are arbitrarily long segments of the number line which contain no primes at all. For example, there can be no prime between $1000! + 2$ and $1000! + 1000$ (why not? [**9.14**])—a run of 999 consecutive non-primes. Can you use this idea to produce a run of a million consecutive natural numbers guaranteed to contain no prime? [**9.15**]

In spite of this, it does appear from the table that the thinning out is remarkably slow. Just look how far we need to go to reduce the primes to one in a hundred.

The question still remains, then: do the primes eventually stop or does an ever more diffuse trace of primes continue along the number line for ever? Before giving Euclid's elegant answer to this question (slightly adapted to show the connection with our proof that $\overline{\mathbb{N}} < \overline{\mathbb{R}}$), we pause to indicate an analogous question about continuous functions. Figures 9.5 and 9.6 show two graphs whose equations are given. You should check that they share the

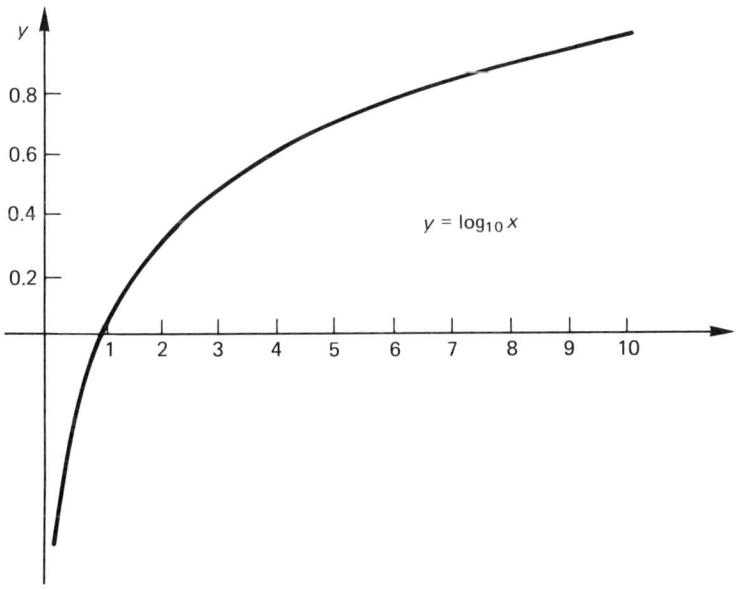

$y = \log_{10} x$

**Fig. 9.5**

$$y = \frac{x-1}{x}$$

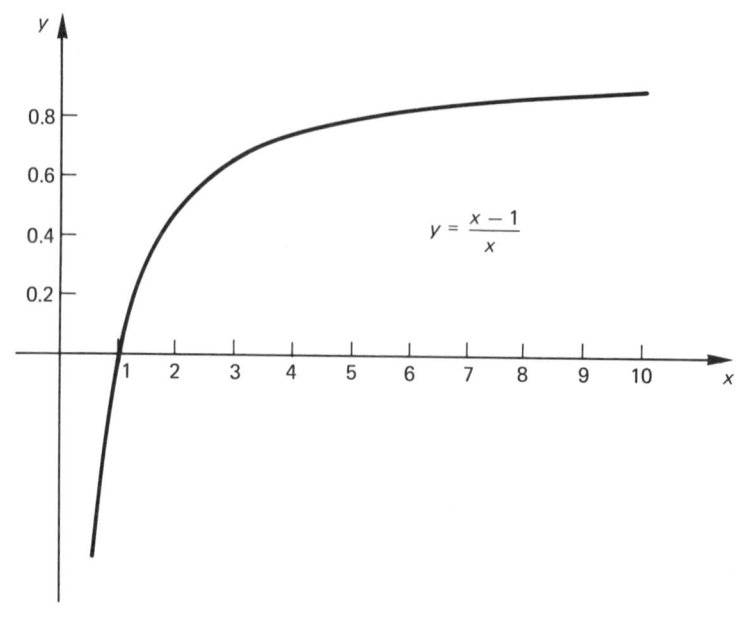

Fig. 9.6

following properties [**9.16**]: $y \to -\infty$ as $x \to 0$ from above; $y = 0$ when $x = 1$; both are increasing functions for $x > 0$; both have continuously decreasing gradients; and both gradients approach zero as $x \to \infty$. But one of them levels off so that $y$ never increases beyond 1, but in the other you can find $x$-values which make $y$ as large as you like. Which is which? [**9.17**]

So, returning to our primes, do they thin out sufficiently fast to die, or do they continue to crop up for ever? Here is our version of Euclid's answer. Take any finite collection $C$ of primes. Multiply them together and add 1 to the result. The number we get can't be divisible by any of the primes in $C$. Why not? [**9.18**] Hence, its prime factors (even if it only has one!) are different from any in $C$. We have shown that no finite set of primes can contain all of them, so the primes must be a denumerable set. We only need to change a couple of words of this to get a summary of what we did in Section 9.3. 'No denumerable set of reals can contain all of them, so $\mathbb{R}$ is bigger than $\mathbb{N}$.' The analogy is now obvious!

*Finding a New Sequence* Figure 9.7 shows a set of six rows of digits. Each row has six digits and each digit is either 1 or 0. Trying to write down a seventh row which is different from the original six could be quite trying by trial and error, but here is a quick method: go down the dotted diagonal and change each digit, so we arrive at 101001. Why is this method bound to work [**9.19**] and what has it to do with Theorem 9.4? We have used the diagonal of an array to obtain a new sequence of digits, and the method of

# SOME INFINITE SURPRISES

```
0  1  0  1  0  0
0  1  1  1  1  0
1  0  0  0  1  1
1  0  1  1  0  0
1  1  0  0  1  1
0  0  1  0  1  0
```
Fig. 9.7

proof of Theorem 9.4 is traditionally referred to as 'the diagonal method'. It is easy to see the reason for the name if we write out the scheme of the proof in the following way.

$r_1$ was characterised by the nest $I_{11}, I_{12}, I_{13}, I_{14} \ldots$
$r_2$ was characterised by the nest $I_{21}, I_{22}, I_{23}, I_{24} \ldots$
$r_3$ was characterised by the nest $I_{31}, I_{32}, I_{33}, I_{34} \ldots$
$r_4$ was characterised by the nest $I_{41}, I_{42}, I_{43}, I_{44} \ldots$

Then the vital number $r$ required in the proof was constructed by manipulating the *diagonal* sequence of intervals $I_{11}, I_{22}, I_{33}, \ldots$.

An identical scheme can be used to prove that the set of all sequences of 0s and 1s is not denumerable.

Here is an arbitrary denumerable set of such sequences:

$$
\begin{array}{cccc}
a_{11} & a_{12} & a_{13} & a_{14} \ldots \\
a_{21} & a_{22} & a_{23} & a_{24} \ldots \\
a_{31} & a_{32} & a_{33} & a_{34} \ldots \\
a_{41} & a_{42} & a_{43} & a_{44} \ldots \\
\vdots & \vdots & \vdots & \vdots
\end{array}
$$

where each $a_{ij}$ is either 0 or 1. Now take the diagonal sequence $a_{11}, a_{22}, a_{33}, \ldots$ and from it produce the sequence $b_1, b_2, b_3, \ldots$, the rule of construction being that if $a_{ii} = 0$, put $b_i = 1$, and if $a_{ii} = 1$, put $b_i = 0$. The result is a sequence which differs from all those in the original list, since it differs from the first in its first place, from the second in its second place, ..., and so on. □

*Bigger and Bigger Infinities* How many subsets does a set have? We found in Chapter 5 that there was an easy answer if the set was finite: an $n$-element set has $2^n$ subsets. This means that finite sets always have more subsets than they have members (almost obvious, though induction provides a very easy formal proof that $2^n > n$, $\forall n \in \mathbb{N}$). What is more surprising is that the same is true of infinite sets, and that we can prove it by using the same strategy as that used to prove that the set of primes is bigger than any finite set, and that $\mathbb{R}$ is bigger than $\mathbb{N}$. That is, we take all the elements of a set $A$, match them up in a one-to-one fashion with subsets of $A$ in any way we like, and show that this process is bound to leave out some of the subsets.

The standard jargon for 'the set of all subsets of a set $A$' is 'the power set of $A$', denoted by $\wp(A)$. So, for example,

$$\wp(\{a,b,c\}) = \{\emptyset, \{a\}, \{b\}, \{c\}, \{b,c\}, \{a,c\}, \{a,b\}, \{a,b,c\}\}$$

In these terms we can state our next theorem.

**Theorem 9.5** For any set $A$, $\overline{\overline{A}} < \overline{\overline{\wp(A)}}$.

*Proof* Let $S_x$ denote the subset of $A$ which we associate with the element $x$ of $A$. The set of all these subsets—that is, $\{S_x : x \in A\}$—is a subset of $\wp(A)$ and what we have to show is that it is not the whole of $\wp(A)$. To do this, we produce a subset $S$ of $A$ which is different from all the $S_x$'s. For each $x \in A$ we ask the question: 'Is $x$ a member of $S_x$?' If the answer is 'no', we put $x$ in $S$; if 'yes', we exclude it from $S$. By this procedure we have ensured that $S$ differs from every $S_x$ and the proof is done. ☐

Figure 9.8 should make clearer what is going on. We have illustrated just seven members of $A$ and their associated seven subsets.

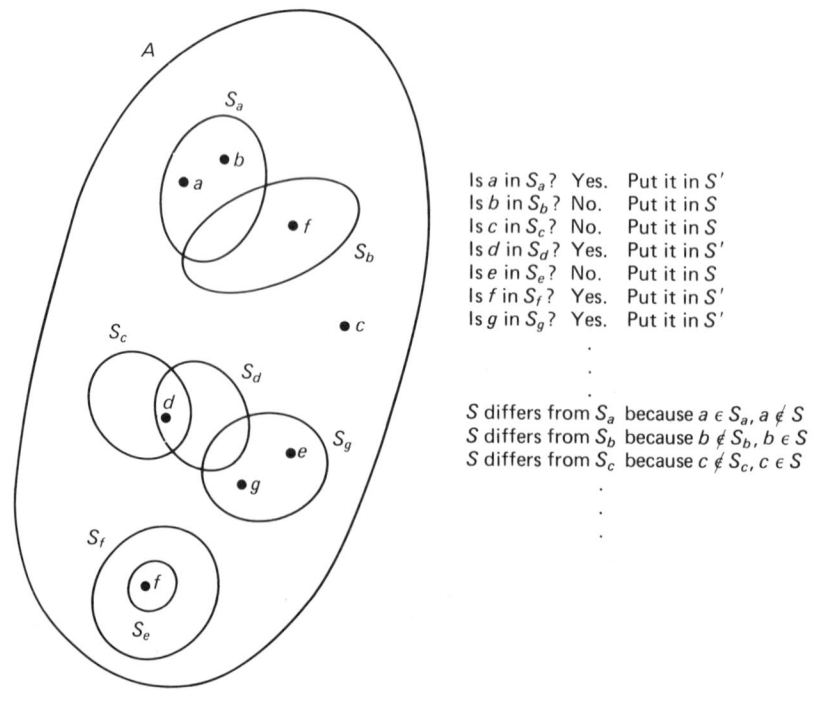

Is $a$ in $S_a$? Yes. Put it in $S'$
Is $b$ in $S_b$? No. Put it in $S$
Is $c$ in $S_c$? No. Put it in $S$
Is $d$ in $S_d$? Yes. Put it in $S'$
Is $e$ in $S_e$? No. Put it in $S$
Is $f$ in $S_f$? Yes. Put it in $S'$
Is $g$ in $S_g$? Yes. Put it in $S'$

$S$ differs from $S_a$ because $a \in S_a$, $a \notin S$
$S$ differs from $S_b$ because $b \notin S_b$, $b \in S$
$S$ differs from $S_c$ because $c \notin S_c$, $c \in S$

**Fig. 9.8**

To show the proof in action for a finite case (say, $A = \{a, b, c\}$), suppose that we decide on $S_a = \emptyset, S_b = \{b, c\}, S_c = \{a, b\}$. Then $S = \{a, c\}$, which is different from $S_a$, $S_b$ and $S_c$.

Our early investigation of infinite sets seemed to indicate that Cantor's method is not a very good size-discriminator between sets. So many unlikely-looking sets all turned out to be the same size as $\mathbb{N}$. But at least we now know from our theorem that there is no limit to bigness in Cantor's sense, since, if you give me any set, I can easily produce a bigger one—namely, its power set.

## 9.5 THE PLANE IS THE SAME SIZE AS THE LINE!

The fact that $\mathbb{R}$ is not countable was proved by Cantor in his first (1874) paper on infinite sets, although for some time before this he believed that $\mathbb{R}$ and $\mathbb{N}$ were equinumerous. This first proof predates by 18 years his invention of the famous diagonal argument and his use of it to prove

$$\overline{\overline{A}} < \overline{\overline{\wp(A)}} < \overline{\overline{\wp(\wp(A))}} < \overline{\overline{\wp(\wp(\wp(A)))}} < \cdots$$

You may, with some justification, feel that the set of all subsets of the set of all subsets of the set of all subsets of $A$ is not exactly a natural object, and that it should be possible to find examples of sets of natural mathematical objects such as numbers, points or functions having greater cardinality than $\mathbb{R}$. Cantor's first attempt to find such a set was a very reasonable, almost obvious, one: $\mathbb{R}$ can be regarded as the set of points of a line, a one-dimensional object, so if we go to two dimensions, $\mathbb{R} \times \mathbb{R}$, the set of all points in the plane, should be considerably larger. He spent three frustrating years from 1874 trying to prove his conjecture that $\overline{\overline{\mathbb{R}}} < \overline{\overline{\mathbb{R} \times \mathbb{R}}}$, and failed for the best possible reason—it just isn't true! At last he became convinced of its falsity and then soon found a proof that $\overline{\overline{\mathbb{R} \times \mathbb{R}}} = \overline{\overline{\mathbb{R}}}$ which he sent to Dedekind in June 1877. The result can be phrased somewhat provocatively as: 'There are just as many points on the line as there are on the plane'. In those days Germans communicated in French, and the letter to Dedekind contained Cantor's famous remark which heads this chapter, 'Je le vois, mais je ne le crois pas'.

Cantor's surprise at this result was due in part to the apparent breakdown of the concept of dimension of a space, and many authorities of the day asserted that it was obvious that 'two independent variables cannot be reduced to one', with the implication that the result must be wrong. Indeed, it was only with the influential help of Weierstrass at this period that Cantor's work was published. The surprising nature of the result is enhanced by the realisation that an easy extension yields a proof that the points of an arbitrarily small line segment can be put into one-to-one correspondence with all the points of a space of $n$ dimensions, or even a space with a denumerably infinite number of dimensions.

It is somewhat ironic that the letter to Dedekind did contain a trivial error, though it was in essence correct. In 1878 he gave a correct proof, not by adapting his original one, but by a method involving continued fractions.

The shock was further compounded in 1890 when Giuseppe Peano managed to define a continuous curve which passed through every point of the unit square. Far from demolishing the idea of dimension, these results led to great activity, and in one direction to a branch of topology still active today—that of Peano spaces. A Peano space is a continuous image of the unit interval.

It was Dedekind who first pointed out the vital role of continuity in the efforts to preserve some sensible meaning for dimension. Cantor's bijection between the unit square and unit interval was hopelessly discontinuous, and Peano's pathological curve, although continuous from the interval to the square, was not one-to-one, so could not be continuous in the other direction. One of the features seized upon to characterise the fundamental difference between square and interval was that there is no function from one onto the other which is continuous in both directions. In modern terminology, this just amounts to the assertion that real continua of dimensions $m$ and $n$ cannot be topologically equivalent unless $m = n$. Strenuous efforts to prove this were successful for dimensions up to 3, mainly owing to the work of Lüroth between 1878 and 1907. Even the case $m = 2$, $n = 3$ is fiendishly difficult, and the general result was not established until 1911, by L. E. J. Brouwer.

Leibniz, almost a century before Cantor, had made a rather cryptic remark (in a philosophical context) to the effect that space constitutes an *order* and not a mere aggregate. The fact that the dimensionality of space is only analysable satisfactorily in terms of continuity seems to be a striking vindication of this view. It also provides a nice counter-example to Whitehead's conjecture that 'whenever a mathematician speaks with misty profundity you can be sure he is talking nonsense'!

Bolzano, too, had made a serious attempt to define dimension about 30 years before Cantor's beginnings of 1874. Unfortunately (perhaps), this was not taken up at the time. Cantor himself published a proof of Brouwer's result in 1879, but it was seriously wrong, which is why the result is still called Brouwer's dimension theorem!

Coming to the present century, in 1927 Sierpinski proved that Cantor's bijection from $[0,1]$ to $[0,1] \times [0,1]$, although not generally continuous, *is* continuous at every *irrational* point. (Chapter 11 will make this strange idea meaningful.) Our final historical snippet on this aspect of Cantor's work is a result of Mardesic proved in 1960. He showed that Peano's curve had hit upon a very special property of real intervals: they are the only spaces $I$ for which there exists a continuous function from $I$ to $I \times I$.

The full story of Cantor's life and work is fascinating, and we direct you to Dauben's book if you wish to learn more. Our main mathematical purpose of this section is to give a proof that the unit square and the unit interval are equinumerous.

Simply to ease the technicalities without sacrificing anything essential, we

# SOME INFINITE SURPRISES

shall consider the case of the half-open interval $I = [0, 1)$ and map this to the half-open square $S = [0, 1) \times [0, 1)$. The latter consists of all points inside and on the unit square, except those points on the right-hand boundary and upper boundary. This is illustrated in Figure 9.9, in which open circles and dotted lines represent parts of the boundary not in $S$. Our task is to find a function $f: I \to S$ which is a bijection.

We need a way of representing real numbers, and the one which is convenient for our task is the usual decimal expansion. We have not mentioned this representation explicitly earlier in the book, but it is just a special case of the monotonic sequence representation mentioned in Chapter 4. For example, our favourite number, $\sqrt{5}$, is $2.236\,067\,977\ldots$, which can also be represented by the convergent monotonic sequence of rational numbers, 2, 2.2, 2.23, 2.236, 2.2360, 2.236 06, .... Every member of $I$ and the coordinates of every member of $S$ will have decimal expansions beginning $0.\ldots$, and since our $f$ is going to be defined by the digits of these decimal expansions, we have to take care that our representation is unique. That is, we don't want more than one decimal to represent the same real number, otherwise we would get entangled in the usual well-defined issue for $f$. The problem arises because a number like 0.42 has many equivalent decimal expansions; 0.42, 0.420, 0.420 00, 0.420 000... and 0.419 999.... Our way round this problem is to deem that whenever a real number has a finite decimal expansion, we represent it by the infinite decimal consisting of the finite expansion followed by an infinite string of zeros. You should check that this is the only way in which ambiguity can arise—in other words, the only reals for which there is a choice of decimal expansion are those which have a finite expansion. [**9.20**]

Now we are ready to define $f$. The essence of the idea is to take each member of $I$—say $a = 0.a_1 a_2 u_3 a_4 \ldots$ and use some of its digits to define the $x$-coordinate of $f(a)$ and the remaining digits to define the $y$-coordinate of $f(a)$. Just for fun, or, if you don't approve of frivolity, for the sound

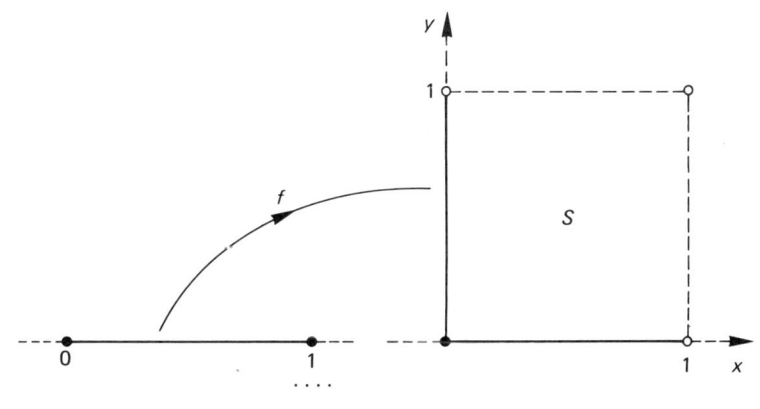

**Fig. 9.9**

educational purpose of keeping you on your toes, the specification of $f$ which we present below will not quite work. We invite you to spot the snag [**9.21**] and, without changing the essential idea above, make a small change in the detail of the definition which will remove the snag. [**9.22**]

The 'nearly right' definition of $f(0.a_1a_2a_3a_4\ldots)$ is $(0.a_1a_3a_5a_7\ldots, 0.a_2a_4a_6a_8\ldots)$.

## 9.6 THE SCHRÖDER–BERNSTEIN THEOREM

We first provide an answer to one of the questions raised about the order relation $<$ in Section 9.2. $<$ has some properties in common with the corresponding relation on $\mathbb{R}$, but there are some important instances in which the two relations have contrasting properties.

Perhaps the most important property shared with the real number order relation is the trichotomy law. For any pair of sets $A$, $B$, one and only one of the following is true: $\bar{\bar{A}} = \bar{\bar{B}}$ or $\bar{\bar{A}} < \bar{\bar{B}}$ or $\bar{\bar{B}} < \bar{\bar{A}}$.

The 'one' part just asserts that any two sets *are* comparable by the order relation, or that $<$ is a *total* order on the class of all cardinals rather than just a *partial* order. Many proofs have been given but they all depend on the theory of ordinal numbers, which is considerably more technical than that of cardinal numbers. [Some of Cantor's early papers assume the unrestricted comparability of cardinals, though in later work he made the assumption explicit. He never succeeded in giving a proof, but he did predict that a proof would demand extremely profound techniques. As predicted, the first proof came in 1904 with Zermelo's well-ordering theorem, tied up with the axiom of choice and all its associated heart-searching.]

The 'only one' part is partially covered by definition in that $\bar{\bar{A}} = \bar{\bar{B}}$ is incompatible with $\bar{\bar{A}} < \bar{\bar{B}}$ and with $\bar{\bar{B}} < \bar{\bar{A}}$. But a nodding acquaintance with the paradoxical habits of infinite sets makes the existence of sets $A$, $B$ with both $\bar{\bar{A}} < \bar{\bar{B}}$ and $\bar{\bar{B}} < \bar{\bar{A}}$ appear quite plausible. Remember that we need only contemplate the possibility of bijections between $A$ and a subset of $B$ and between $B$ and a subset of $A$, but *no* bijection between $A$ and $B$. That such contemplation is futile is the essence of the famous Schröder–Bernstein theorem, named after Bernstein for getting it right, and Schröder for getting it wrong, independently, in the same year, 1897!

One way of stating the theorem is $(\bar{\bar{A}} \leqslant \bar{\bar{B}}$ and $\bar{\bar{B}} \leqslant \bar{\bar{A}}) \Rightarrow (\bar{\bar{A}} = \bar{\bar{B}})$, and it is a result of great significance for the following reason. It is often quite easy to establish a bijection from $A$ to a subset of $B$ and vice versa, but very difficult to give a direct demonstration of a bijection from $A$ to $B$. The Schröder–Bernstein theorem just says that we don't need to do the hard bit, because it is implied by the easy bit!

## 9.6.1 Proof of the Schröder–Bernstein Theorem

> 'There seems to be a principle of conservation of difficulties which says that a hard theorem is hard, no matter how you approach it.'
> Ian Richards, in *Mathematics Today* (Springer, 1978)

There are several proofs, but, in view of the quotation above, we need make no apology for the Schröder–Bernstein theorem being a tough one. Surprisingly tough, given that the corresponding result for finite sets is a triviality.

Sir Ernest Shackleton said, in rather a different context, that 'difficulties are just things to be overcome', so if you are feeling mentally sufficiently frisky to join him, the proof presented below is well worth studying. If not, skip it. There is no penalty!

*Preliminary Concepts* (1) *The vital lemma*  To make a start, consider the function $w$, which is a function from the power set of $\{1,2,3\}$ into itself. We define it directly by listing all its images:

$w(\emptyset) = \{2\}$; $w(\{1\}) = \{2,3\}$; $w(\{2\}) = \{2,3\}$; $w(\{3\}) = \{1,2\}$;
$w(\{1,2\}) = \{2,3\}$; $w(\{1,3\}) = \{1,2,3\}$; $w(\{2,3\}) = \{1,2,3\}$;
$w(\{1,2,3\}) = \{1,2,3\}$

Notice that $w$ has a 'fixed set' $\Omega = \{1,2,3\}$ in the sense that $w(\Omega) = \Omega$. It also has the property, not quite so obvious but easy enough to check, of being inclusion-preserving. That is, if $X \subseteq Y$, then $w(X) \subseteq w(Y)$. These two properties are related, and our proof hinges on the following lemma, which expresses this relationship.

---

**Lemma**  If $A$ is a set and $w: \wp(A) \to \wp(A)$ is inclusion-preserving, then $w$ has a 'fixed set' $\Omega$ such that $w(\Omega) = \Omega$.

---

We delay proof of this and go on to:

(2) *A convenient abuse of notation*  If $f: \mathbb{R} \to \mathbb{R}$ is the sine function, then $f(0)$, $f(\pi/2)$, $f(3)$, etc., all make sense, but $f([0,1])$, $f(\mathbb{R})$ are meaningless, because $[0,1]$ and $\mathbb{R}$ are not real numbers but *sets* of real numbers. However, it seems reasonable to say that if $S$ is a set of reals, then $f(S)$ can be taken to mean the set of their sines. For example, if

$$S = \left\{0, \pi, \frac{\pi}{2}, \frac{3\pi}{2}, \frac{5\pi}{2}, \frac{\pi}{4}\right\}$$

then

$$f(S) = \left\{0, 1, -1, \frac{1}{\sqrt{2}}\right\}$$

It is, according to the strict letter of mathematical law, illegal, because this new $f$ is a function from $\wp(\mathbb{R})$ to $\wp(\mathbb{R})$ and we should not use the same letter, $f$, to denote different things. We shall condone the abuse and rely on the context to avoid ambiguity.

*The Main Proof* Suppose that $A_s$ and $B_s$ are subsets of $A$ and $B$, respectively, and we have bijections $f: A \to B_s$ and $g: B \to A_s$ as in Figure 9.10. For the next stage in the argument, we abuse notation as in note (2) above and regard $f$ and $g$ as functions $\wp(A) \to \wp(B)$ and $\wp(B) \to \wp(A)$, respectively.

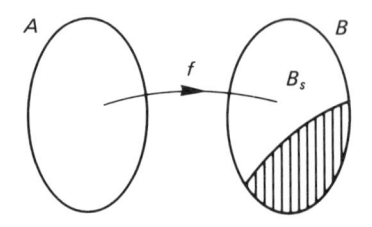

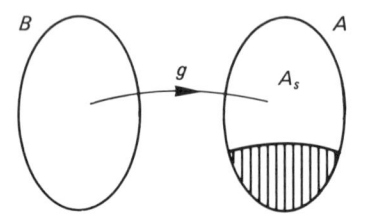

**Fig. 9.10**

For each $X \subseteq A$, we define the set $X^* \subseteq A$ by $X^* = A \setminus g(B \setminus f(X))$. This strange convoluted definition of $X^*$ is simply a device to enable us to use the lemma. Note that, given any subset $X$ of $A$, the corresponding $X^*$ is also a subset of $A$, so the map which gets us from $X$ to $X^*$ is a map from $\wp(A)$ to $\wp(A)$. We now show that it has the inclusion-preserving property by taking $X_0, X_1 \in \wp(A)$ with $X_0 \subseteq X_1$ and deducing that $X_0^* \subseteq X_1^*$.

[9.23] Check that you can justify each step below:

$$X_0 \subseteq X \Rightarrow f(X_0) \subseteq f(X_1)$$
$$\Rightarrow B \setminus f(X_0) \supseteq B \setminus f(X_1)$$
$$\Rightarrow g(B \setminus f X_0)) \supseteq g(B \setminus f(X_1))$$
$$\Rightarrow A \setminus g(B \setminus f(X_0)) \subseteq A \setminus g(B \setminus f(X_1))$$

i.e.

$$X_0^* \subseteq X_1^*$$

The lemma now enables us to assert the existence of a 'fixed set' $\Omega$ of the * map; that is, $\Omega = \Omega^*$. In other words,

$$\Omega = A \setminus g(B \setminus f(\Omega))$$

Hence,

$$A \setminus \Omega = g(B \setminus f(\Omega))$$

[9.24] Check this last deduction by convincing yourself that for any sets $P, Q, P \setminus (P \setminus Q) = Q$.

# SOME INFINITE SURPRISES

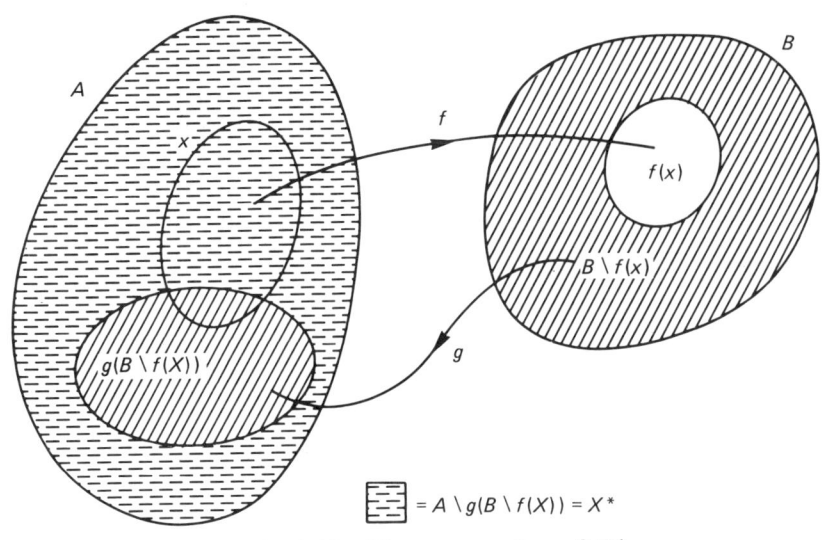

**Fig. 9.11** The construction of $X^*$

Now at last we can define our bijection from $A$ onto $B$. It is done by the formula:

$$h(x) = \begin{cases} f(x) & \text{if } x \in \Omega \\ g^{-1}(x) & \text{if } x \in A \setminus \Omega \end{cases}$$

and is illustrated in Figure 9.12.

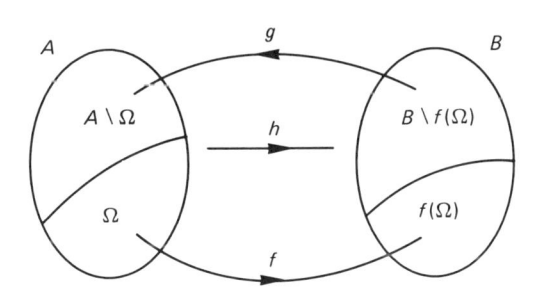

**Fig. 9.12** The construction of $h$ from $f$, $g$ and $\Omega$

[9.25] Now polish off the proof by checking that this $h$ really is a bijection from $A$ onto $B$. You have to be sure that no two elements of $A$ can have the same image in $B$, that each element of $A$ does have an image in $B$, and that each element of $B$ is the image of something in $A$.

*Proof of the Lemma* We have delayed this, not just to avoid disturbing the flow but because the lemma is really a very nice result in its own right, and it seems somewhat unjust to relegate it to the status of a mere lemma.

Let $w$ be an inclusion-preserving map on $\wp(A)$. Now, if $X$ is a subset of $A$ (that is, $X \in \wp(A)$), then $X$ may or may not be included in its $w$-image, $w(X)$. We consider the collection $S$ of all subsets which are so included. Formally,

$$S = \{X : X \in \wp(A), X \subseteq w(X)\}$$

[**9.26**] What is $S$ for the example on $\{1, 2, 3\}$?

For $\Omega$ we take the union of all the subsets in $S$:

$$\Omega = \bigcup_{X \in S} X$$

[**9.27**] What is $\Omega$ for the same example?

We claim that $\Omega$ is the required fixed set for $w$. Take any $X$ in $S$ so that, using the defining property of $S$,

$$X \subseteq w(X) \tag{1}$$

But also $X \subseteq \Omega$ by our definition of $\Omega$. It follows, by the inclusion-preserving property of $w$, that

$$w(X) \subseteq w(\Omega) \tag{2}$$

Putting (1) and (2) together, we deduce $X \subseteq w(\Omega)$.

But $X$ was an arbitrary member of $S$, so what we have proved is that *every* $X$ in $S$ is a subset of $w(\Omega)$. Hence, their union is a subset of $w(\Omega)$. That is, $\Omega \subseteq w(\Omega)$, and we are half-way to proving our claim. The other half is, of course, to show that $w(\Omega) \subseteq \Omega$. This is incredibly simple and neat: simple because all we do is apply, in turn, the inclusion-preservation of $w$, the definition of $S$ and the definition of $\Omega$; neat because we make $\Omega \subseteq w(\Omega)$ imply its own reversal. It is all done in three lines:

$$\Omega \subseteq w(\Omega) \Rightarrow w(\Omega) \subseteq w(w(\Omega)) \text{—property of } w$$
$$\Rightarrow w(\Omega) \in S \qquad \text{—definition of } S$$
$$\Rightarrow w(\Omega) \subseteq \Omega \qquad \text{—definition of } \Omega$$

Done! □

*Remark* There are several reasons for being fond of this lemma. First, it sounds so unlikely—taking into account the unlimited size of $A$ and the innumerable choices of inclusion-preserving $w$'s, it seems incredible that such a weak requirement as inclusion preservation is nevertheless strong enough to ensure the existence of $\Omega$. Second, the proof is exciting because of its simplicity—just repeated use of the meanings of $\cup$, $\subseteq$ and $\in$. Third, its ingenuity—spotting the $\Omega$ which will do the trick. Actually, this production of $\Omega$ is not quite the rabbit out of a hat that it seems; there are quite a lot of results in algebra and topology where we obtain a set with a required property by looking at intersections or unions of all sets with a similar property (recall that $\Omega$ was the union of all sets in $S$). So spotting $\Omega$ is a matter of experience as much as ingenuity, but this only adds to the interest of the proof rather than detracts from it.

The lemma passes the G. H. Hardy test with distinction: 'Beauty is the first test: there is no permanent place in the world for ugly mathematics.'

## 9.6.2 An Application of the Schröder–Bernstein Theorem

To display the power of the theorem, we shall prove that the set $P$ of all permutations of $\mathbb{N}$ is equinumerous with the set of all real numbers in the open unit interval $(0, 1)$.

As we did previously, we shall represent the real numbers by their decimal expansions, ensuring uniqueness of representation by agreeing that any decimal which would normally terminate is to be written in the version involving an infinite string of zeros.

*Construction of a Bijection from* $P$ *onto a Subset of* $(0, 1)$  Suppose that a permutation of $\mathbb{N}$ begins $3, 10, 94, 7, \ldots$. We map this to the real number

$$0.\underbrace{111}_{3\ 1s}\ \underbrace{2\,2\ldots2}_{10\ 2s}\ \underbrace{1\,1\ldots1}_{94\ 1s}\ \underbrace{2\,2\ldots2}_{7\ 2s}\ 1\ldots$$

so our bijection is from $P$ onto the set of reals in $(0, 1)$ whose decimal expansions consist of alternating blocks of 1s and 2s (1s first) such that every block length appears once and once only—a very 'small' subset of $(0, 1)$!

*Construction of a Bijection from* $(0, 1)$ *onto a Subset of* $P$  Given the real number $r$, consider its (unique) decimal representation, $0.d_1 d_2 d_3 \ldots$. Add 2 to each of its digits, to obtain the sequence $e_1, e_2, e_3, \ldots$, where $e_i = d_i + 2$. From this generate the following permutation of $\mathbb{N}$: put the first $e_1$ natural numbers in their natural order, the next $e_2$ in reverse order, the next $e_3$ in natural order..., and so on. For example, from $r = 0.31500704\ldots$ we get $e_1 = 5$, $e_2 = 3$, $e_3 = 7$, $e_4 = 2$, $e_5 = 2$, $e_6 = 9$, $e_7 = 2$, $e_8 = 6\ldots$ and the corresponding permutation of $\mathbb{N}$ is

$$\underline{1\ 2\ 3\ 4\ 5}\ \underline{8\ 7\ 6}\ \underline{9\ 10\ 11\ 12\ 13\ 14\ 15}\ \underline{17\ 16}\ \underline{18\ 19}$$
$$\underline{28\ 27\ 26\ 25\ 24\ 23\ 22\ 21\ 20}\ \underline{29\ 30}\ \underline{36\ 35\ 34\ 33\ 32\ 31}\ \ldots$$

Again the bijection is onto a very 'small' special subset of $P$, but still the Schröder–Bernstein theorem assures us that we have done enough to guarantee the existence of a bijection from *all* of $(0, 1)$ to *all* of $P$. □

[9.28] Why is the 'add 2' step essential? (Two reasons.)

Suppose that we didn't have the S–B theorem available and the only way of showing that $(0, 1)$ and $P$ were equinumerous was to define an explicit bijection between them. Contemplate this task. It should encourage great respect for the theorem!

## 9.7 A TOTALLY DIFFERENT WAY OF MEASURING INFINITY

Although $\mathbb{R}$ and its subset $\mathbb{Q}$ are both infinite, densely packed sets, you should now have some feeling for the former being 'infinitely more infinite' than the latter, and that this notion can be made precise, using the language of bijections.

This section is concerned with another way of capturing the relative smallness of the rationals. Just to keep things finite (in length, not number!), we concentrate on $I$, the unit real interval $(0, 1)$, and $I_\mathbb{Q}$, its subset consisting of just its rational members.

What we are after is a way of measuring the 'length' of $I_\mathbb{Q}$, and we hope the answer will be smaller than 1, the length of $I$, thus giving more precise reinforcement to our knowledge that $I_\mathbb{Q}$ is 'smaller' than $I$.

One idea is to ask 'what is the smallest interval which will cover $I_\mathbb{Q}$?' and use its length as the 'length' of $I_\mathbb{Q}$. Unfortunately, the answer is $I$, so this fails to distinguish between $I$ and $I_\mathbb{Q}$ and we must look for something more refined. Our first attempt would also fail to distinguish between $I$ and $(0, 0.1) \cup (0.9, 1)$, the answer being $I$, of length 1 in both cases, whereas any sensible definition of the length of a set should give 0.2 for the second one. However, this observation does give us a clue—don't stop at single intervals.

Following this line for $I_\mathbb{Q}$, we know that $I_\mathbb{Q}$ is denumerable, so we can envisage a listing of its members $q_1, q_2, q_3, \ldots$. If we cover $q_1$ with an interval of length $\frac{1}{4}$, $q_2$ with one of length $\frac{1}{8}$, $q_3$ with one of length $\frac{1}{16}$, and so on, the total length of all these intervals is $\frac{1}{4} + \frac{1}{8} + \frac{1}{16} + \frac{1}{32} + \cdots = \frac{1}{2}$. (We are anticipating Chapter 10 here, but even if you don't know about convergence of series in general, you should be able to convince yourself that the sequence $\frac{1}{4}, \frac{1}{4} + \frac{1}{8}, \frac{1}{4} + \frac{1}{8} + \frac{1}{16}, \ldots$ has a limit of $\frac{1}{2}$.)

So the length of $I_\mathbb{Q}$ is at most $\frac{1}{2}$ ('at most') (a) because we have allowed our covering intervals to overlap, and (b) because we could clearly get a smaller total by starting with a shorter interval or by doing something more drastic than halving the length at each step). Pushing this idea further, we see that the total can be made *as small as we like*—to get a total of $10^{-6}$, for example, all you need to do is cover $q_1, q_2, q_3, \ldots$ by intervals of lengths $10^{-6}/2, 10^{-6}/4, 10^{-6}/8, \ldots$. The outcome of this limiting argument is that it makes some sort of sense to say that $I_\mathbb{Q}$ has zero length. This is satisfying in the sense that if $I_\mathbb{Q} \subseteq I$ and $\bar{I}_\mathbb{Q} < \bar{I}$, it is reasonable that $I_\mathbb{Q}$ should be 'shorter' than $I$ too. The trouble is that the same argument which assigned zero length to $I_{\mathbb{Q}}$ will also assign zero length to any countable subset of $\mathbb{R}$, so nothing much seems to have been gained.

Don't feel let down by this. It would be unkind, even immoral, to lead you up this severe route only to reward you with anticlimax. We are not immoral, so we close this rather long chapter with a truly mind-boggling set $\mathscr{A}$. It, too, is a subset of $I$. Its points are so sparsely spread over $I$ that it seems minute compared with $I_{\mathbb{Q}}$, so it should come as no surprise that its length is zero. But in terms of cardinality it is exactly the same size as $I$!

### 9.7.1 The Construction of $\mathscr{A}$

Start with $I$. Throw out its middle closed third, $[\frac{1}{3},\frac{2}{3}]$. This leaves the set $A_1 = (0,\frac{1}{3}) \cup (\frac{2}{3},1)$. Now throw out the middle closed third of each interval making up $A_1$. This leaves $A_2 = (0,\frac{1}{9}) \cup (\frac{2}{9},\frac{1}{3}) \cup (\frac{2}{3},\frac{7}{9}) \cup (\frac{8}{9},1)$. Carry on in this way to produce $A_3, A_4, \ldots$. The first four steps are shown in Figure 9.13. $\mathscr{A}$ is just what is ultimately left. Not a lot, you may think. Indeed, you may be forgiven for thinking that $\mathscr{A} = \emptyset$, but here intuition lets us down badly. One thing we can be sure of is that the length of $\mathscr{A}$ must be zero, because if we add up all the lengths of the bits taken out of $I$ to make $\mathscr{A}$, we get

$$\frac{1}{3} + 2 \times \frac{1}{9} + 4 \times \frac{1}{27} + 8 \times \frac{1}{81} + 16 \times \frac{1}{243} + \cdots$$
$$= \frac{1}{3}[1 + \frac{2}{3} + (\frac{2}{3})^2 + (\frac{2}{3})^3 + (\frac{2}{3})^4 + \cdots]$$
$$= 1$$

because the square brackets contain a geometric series whose limit is 3.

So the length of $\mathscr{A}$ is zero. How can it possibly be equinumerous with $I$? To answer this, we use the tricimal representation of real numbers. This just amounts to using 3 as our base instead of 10. Just as the decimals 734.302 and 0.$\dot{3}$ mean

$$7 \times 10^2 + 3 \times 10^1 + 4 \times 10^0 + 3 \times 10^{-1} + 0 \times 10^{-2} + 2 \times 10^{-3}.$$

and

$$3 \times 10^{-1} + 3 \times 10^{-2} + 3 \times 10^{-3} + \cdots$$

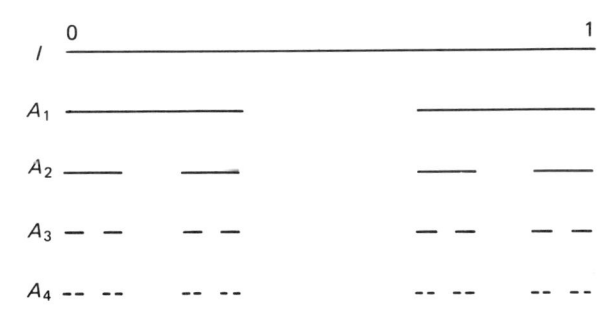

**Fig. 9.13** The early stages in the construction of $\mathscr{A}$

respectively, so the tricimals 201.21 and 0.$\dot{2}$ mean

$$2 \times 3^2 + 0 \times 3^1 + 1 \times 3^0 + 2 \times 3^{-1} + 1 \times 3^{-2}$$

and

$$2 \times 3^1 + 2 \times 3^{-2} + 2 \times 3^{-3} + \cdots$$

respectively. Note that any real number has a tricimal representation and that tricimals only make use of the digits 0, 1 and 2.

Now go back to our method of constructing $\mathscr{A}$. $I$ just contains all the tricimals (except zero) beginning 0. . . . . To obtain $A_1$ we threw out all those tricimals representable with a 1 in the first tricimal place. $A_2$ is formed by removing all remaining tricimals with a 1 in the second place, . . . , and so on. After $n$ steps all members remaining in $A_n$ have no 1 in their first $n$ tricimal places, and it is *only* those reals in $I$ with a 1 somewhere in the first $n$ places which have been excluded.

Ultimately we are left with $\mathscr{A}$, which must contain exactly those members of $I$ whose tricimals contain only 0s and 2s. We leave you with the final exercise:

[**9.29**] Use the above characterisation of $\mathscr{A}$ to prove that $\bar{\bar{\mathscr{A}}} = \bar{\bar{I}}$. [*Hint*: The Schröder–Bernstein theorem is indispensable!]

## 9.8 WHERE NEXT?

One topic we have not had space to mention is the arithmetic of cardinal numbers. We can give meanings to the sum or product of arbitrarily large infinite collections of arbitrarily large infinite cardinal numbers. Furthermore, this can be done in such a way that they obey the ordinary rules such as

$$a(b+c) = ab + ac; \quad (a^b)^c = a^{bc}; \quad (ab)^c = a^c b^c$$

but some of their other properties are truly bizarre. Their arithmetic is also extremely powerful, delivering, for example, a one-line proof that the infinite dimensional space $\mathbb{R}^{\mathbb{N}}$ is equinumerous with $(0, 1)$. But there we must, regretfully, leave it with a reference to a book which though old is still a classic: *Stories about Sets*, by N. Ya Vilenkin.

# 10 Sequences and Series—in which we discover very odd behaviour in even the smallest infinite set

'A quantity which is increased or decreased by an infinitely small quantity is neither increased nor decreased.'

Johann Bernouilli

'How often have I said to you that when you have eliminated the impossible, whatever remains, however improbable, must be the truth.'

Sir Arthur Conan Doyle

## 10.1 SEQUENCES

Much of the effort we have expended in developing the system $\mathbb{R}$ of real numbers and in describing the infinite can now be put to good use. In this chapter we shall be investigating limiting processes, the very foundation of analysis. We begin by agreeing that an *infinite sequence* is a countably infinite set of real numbers occurring in some definite order, $a_1, a_2, a_3, \ldots, a_n, \ldots$. Each $a_i \in \mathbb{R}$ and there is one $a_i$ for each $i \in \mathbb{N}$. A favoured abbreviation for a sequence is $(a_n)$, where $a_n$ denotes the $n$th term of the sequence.

The reader is perhaps familiar with questions (much used in so-called intelligence tests) such as: What is the next number in the sequence 1, 4, 9, 16,...? The answer sought is 25. The question is, of course, silly, since it assumes that one recognises the sequence as $(n^2)$. It could, of course, be the sequence $(n^4 - 10n^3 + 36n^2 - 50n + 24)$, in which case the next term is $-1051$! Hence, we must specify a sequence by indicating, in some way, the precise method by which successive terms of the sequence are generated. This is most easily done by giving a formula, involving $n$, for the $n$th term. However, we could have any description which is theoretically capable of uniquely determining the $n$th term — for example, the sequence whose $n$th term is the $n$th decimal digit in the decimal expansion of $\pi$. Any algorithm for producing 'the next term' from previous ones will also suffice.

### 10.1.1 Examples

(1) The Fibonacci sequence (seen in Chapter 5) is given by $a_1 = 1$, $a_2 = 1$

and $a_{n+2} = a_{n+1} + a_n$ for all $n \geq 1$. The first few terms of this sequence are 1, 1, 2, 3, 5, 8, 13,.... There is a formula for the $n$th term of the Fibonacci sequence, which is that

$$a_n = ((1 + \sqrt{5})^n - (1 - \sqrt{5})^n)/2^n \sqrt{5}$$

This can be established by induction on $n$ [**10.1**], although it is somewhat harder to imagine how it might be derived from scratch. The sequence arises as a consequence of one of the problems posed in the *Liber abaci*, a work published by Leonardo Fibonacci in 1202. The problem in question is as follows: How many pairs of rabbits can be produced from a single pair in a year if each month each pair begets a new pair which from the second month on becomes productive?

(2) By simple algebra we can rewrite the equation $x^2 = 5$ rather arbitrarily as $x = (x^2 + 5)/2x$. Now we define a sequence by $a_1 = 3$ and

$$a_{n+1} = (a_n^2 + 5)/2a_n \quad \text{for} \quad n \geq 1$$

The first few terms in this sequence are 3, 2.33, 2.2381, 2.236 069, 2.236 068. It is amazing, but true, that successive terms of this sequence give better and better approximations to $\sqrt{5}$. We shall prove this fact later in this chapter.

(3) In contrast to our previous example, consider the sequence given by $a_1 = 3$ and $a_{n+1} = 5/a_n$ for $n \geq 1$. (Suggested, of course, by rewriting $x^2 = 5$ as $x = 5/x$.) It is fairly easy to see that this sequence oscillates between 3 and 5/3 for evermore.

(4) If $(a_n)$ denotes the Fibonacci sequence of example (1), then define a sequence $(b_n)$ as follows:

$$b_n = a_{n+1}/a_n \quad \text{for} \quad n \geq 1$$

This sequence proceeds as

$$1, 2, 3/2, 5/3, 8/5, 13/8, \ldots$$

or, in decimal form,

$$1, 2, 1.5, 1.6666\ldots, 1.6, 1.625, \ldots$$

The terms of this sequence approach, ever more closely, a value known as the golden ratio. We shall return to this example in due course.

We can see from these examples that it is often not at all obvious what the general term of a given sequence is. Also, many different types of behaviour are apparent. The terms of the Fibonacci sequence increase without bound. In example (3) the terms oscillate between two different values. In examples (2) and (4) a more settled behaviour can be seen. In both cases the terms of the sequence appear to be converging on some fixed real number—what exactly this limiting value is is not at all clear. Before attempting a formal definition of what we shall call a *convergent sequence*, we consider a few more examples which we hope show that our intuition cannot always be relied on to spot either convergence or even the correct limit of a convergent sequence. We feel that this is a worth-while exercise, since it should have you screaming out for the official definition of convergence.

(5) Let
$$a_n = (1/n^2 + 2/n^2 + 3/n^2 + \cdots + n/n^2) \quad \text{for each} \quad n \geq 1$$
$$a_n = (1 + 2 + 3 + \cdots + n)/n^2 = \tfrac{1}{2}n(n+1)/n^2 = \tfrac{1}{2}(1 + 1/n)$$

Now as $n$ increases (which we shall denote for the time being by $n \to \infty$) the term $1/n$ becomes smaller and smaller (intuitively obvious!). So the terms of the sequence appear to decrease steadily in size and approach the value $\tfrac{1}{2}$. In other words, we feel intuitively that $(a_n)$ is converging to $\tfrac{1}{2}$. However, for any fixed natural number $k$, $1 \leq k \leq n$, the term $k/n^2$ in the original expression for $a_n$ rapidly decreases towards zero as $n \to \infty$ (again, intuitively obvious!?) Hence, all of $1/n^2, 2/n^2, \ldots, n/n^2$ decrease towards zero as $n \to \infty$, so surely their sum $a_n$ also decreases to zero. This intuition cannot be sound, for how can $(a_n)$ converge to both 0 and $\tfrac{1}{2}$? The first argument is correct: $(a_n)$ does converge to $\tfrac{1}{2}$, so explain why the second argument is false. [**10.2**]

(6) Let $a_n = \sqrt{(n+1)} - \sqrt{n}$ for $n \geq 1$. Now
$$a_n = (\sqrt{(n+1)} - \sqrt{n})(\sqrt{(n+1)} + \sqrt{n})/(\sqrt{(n+1)} + \sqrt{n})$$
$$= 1/(\sqrt{(n+1)} + \sqrt{n})$$

Now as $n$ increases, the denominator of $a_n$ increases, we feel, without bound and so a good guess is that $(a_n)$ converges to zero. But why the fancy algebra? Surely a simpler intuitive argument can be used along the following lines. For large values of $n$, $n+1$ is approximately equal to $n$ and so for large values of $n$, $a_n$ is approximately zero. Neither argument, at this stage, can be properly called a proof. There is too much imprecision with notions such as 'When $n$ is large, such and such is small'. Your attention is drawn to Bernouilli's quote at the beginning of this chapter.

(7) Shades of example (6) appear again in this example, where now
$$a_n = \sqrt{(n(n+1))} - n \quad \text{for each} \quad n \geq 1$$

We could give some credence to the following argument. When $n$ is very large, $n^2$ is huge, and for large enough $n$, $n^2 + n$ is approximately $n^2$. Hence, $a_n$ is approximately zero (for example, $a_{100} = 0.4988$). However, you are invited to attempt some clever algebra as we did in example (6) to deduce that in this example $(a_n)$ actually appears to converge to $\tfrac{1}{2}$. [**10.3**]

(8) Refer to Figure 10.1, where we have a semicircle of radius $AO = OB = 1$ and successively we have constructed 2, 4, 8, 16, ... semicircles along AB, as indicated. We define a sequence $(a_n)$ by letting $a_n$ be the total arc length of the $2^n$ semicircles at the $n$th stage of our construction. When there are $2^n$ semicircles along AB, each one has radius $2^{-n}$. The arc length of a semicircle of radius $2^{-n}$ is $\pi \cdot 2^{-n}$, and since there are $2^n$ such semicircles, $a_n = 2^n \cdot \pi \cdot 2^{-n} = \pi$. Our sequence is the constant sequence $\pi, \pi, \pi, \ldots$. Any sensible definition of convergence would surely give $\pi$ as the limit of this

[Fig. 10.1]

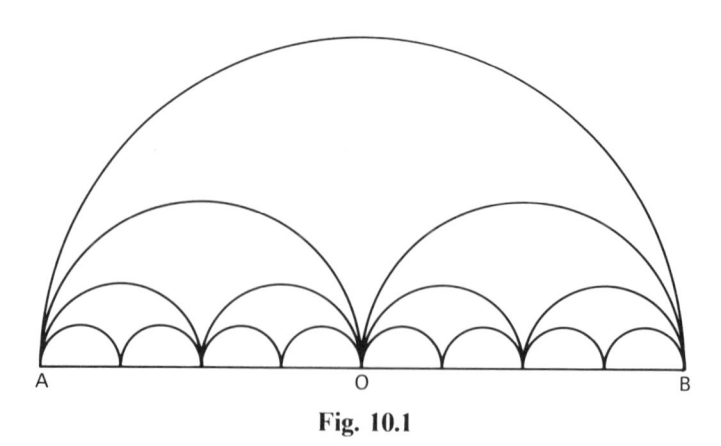

**Fig. 10.1**

sequence—and it does! But the paradox here, which we shall raise again in Chapter 11, is that, as $n$ increases, the path from A to B along the semicircles appears to get closer to the straight-line path from A to B. But $AB = 2$ and so $\pi$ equals 2.

## 10.2 CONVERGENCE

Let us now work towards an official definition of a sequence $(a_n)$ tending to a limiting value $L$ as $n$ tends to infinity. We shall denote the convergence of such a sequence $(a_n)$ by writing either $\lim_{n \to \infty}(a_n) = L$, or $(a_n) \to L$ as $n \to \infty$, or even more simply as $(a_n) \to L$. To make life a little easier, we consider first a sequence $(A_n)$ whose limiting value will turn out to be zero. In this case, as we proceed along the sequence $A_1, A_2, A_3, \ldots, A_n, A_{n+1}, \ldots$, we require that eventually the terms of the sequence are small. For example, there must be some point of the sequence, say $A_N$, where *all* the terms after $A_N$ are all in magnitude less than $10^{-6}$. Likewise, there must be some term of the sequence—$A_M$, say—where all successive terms have size less than $10^{-12}$. $M$ will no doubt be a much larger natural number than $N$. But now we need some $A_R$, where all later terms have absolute value less than $(\text{googol})^{-1}$. What on earth is a googol, you ask! 'A googol is so large that you cannot name it or talk about it; it is so large that it is infinite.' This lay definition of a very large number is superseded by the definition that 1 googol = $10^{100}$ in Edward Kasner's essay entitled 'New names for old'. (See J. R. Newman's *The World of Mathematics*, Simon and Schuster, 1959–60, p. 2009.) The name was invented by Kasner's nine-year-old nephew, who also coined the name googolplex for a 1 followed by as many 0s as you can write before getting tired. The point we are making is that, in order for $(A_N)$ to have limit zero, we must be able to ensure that from some point onwards all the terms are less in magnitude than some prescribed positive number. This number, or tolerance, which is not under our control, may be incredibly small. This, of course, is the whole point of the exercise.

# SEQUENCES AND SERIES

**Example** Consider $A_n = \cos(n)/2^n$ for $n \geq 1$.
$A_1 = 0.27$, $A_2 = -0.10$, $A_3 = -0.12$, $A_4 = -0.04$, $A_5 = 0.01,\ldots$. By comparing powers of 2 with powers of 10 and using the fact that $|A_n| \leq 2^{-n}$ [**10.4**], we can show that the tolerances prescribed in the table of Figure 10.2 require the indicated $N$'s. In each case $|A_n| < \varepsilon$ for all $n$ exceeding $N$.

| $\varepsilon$ | $N$ |
|---|---|
| $10^{-6}$ | 19 |
| $10^{-12}$ | 39 |
| (googol)$^{-1}$ | 332 |

**Fig. 10.2**

Now if we can manufacture an $N$, depending on $\varepsilon$, such that $|A_n| < \varepsilon$ for all $n$ exceeding $N$ (no matter what small $\varepsilon$ is given), then the terms of the sequence are forced to get arbitrarily close to 0 eventually! Since $|A_n| \leq 2^{-n}$ and $2^{-n} < \varepsilon$ if and only if $n > \log_2(1/\varepsilon)$ [**10.5**], then the $N$ required is any integer exceeding $\log_2(1/\varepsilon)$.

---

**Official Definition of Convergence to Zero** A sequence $(a_n) \to 0$ as $n \to \infty$ if and only if for each $\varepsilon > 0 \;\exists$ a natural number $N = N(\varepsilon)$ such that $n > N \Rightarrow |a_n| < \varepsilon$.

---

Note that in our preceding example we proved that $(\cos n)/2^n \to 0$.

The official definition of convergence to an arbitrary real number $L$ is now easy to imagine. For $(a_n)$ to have limit $L$, we simply demand that the sequence $(a_n - L)$ have limit 0.

---

**Official Definition of Convergence** A sequence $(a_n) \to L$ as $n \to \infty$ if and only if for each $\varepsilon > 0 \;\exists$ a natural number $N = N(\varepsilon)$ such that $n > N \Rightarrow |a_n - L| < \varepsilon$.

---

Provided that a given sequence converges and that we either know or can guess what its limit $L$ must be, then we can prove that it does indeed converge to $L$ by implementing the definition above. In other words, for each $\varepsilon > 0$ we have to produce the requisite $N(\varepsilon)$. The value of our official definition is that it does not depend on individual intuition. In fact, the definition underpins intuitive ideas when they are sound. It is not our purpose here to pursue an analysis course but merely to whet your appetite for one. Suffice it to say that this definition of $\lim_{n \to \infty}(a_n) = L$ can be (and is) used to establish rigorously the convergence of certain simple sequences. It is not much use for complicated sequences, and is, of course, useless if we have no idea of what $L$, if it exists, might be. The definition is thus used to establish certain theoretical results which allow us to algebraically manipulate sequences in order to 'read off' the limit. As an example of these processes, we return to example (7) of Section 10.1. We have that

$$(a_n) = \sqrt{(n(n+1))} - n = n/(\sqrt{(n(n+1))} + n) = 1/(\sqrt{(1+1/n)} + 1)$$

Now, using the fact that $1/n \to 0$ (easily shown by our official definition [**10.6**]) and that a constant sequence of 1's has limit 1 (can you prove this? [**10.7**]), the theoretical results alluded to above guarantee that $(a_n) \to 1/(\sqrt{(1+0)} + 1) = 1/2$ as $n \to \infty$.

Not all sequences are amenable to such manipulations and a more fundamental question arises—given a sequence with no obvious limit, how can we decide whether or not it converges? We require some tests for convergence. These are available, provided that we are prepared to take on board more concepts. Once again, as so often in mathematics, new definitions, born out of necessity, loom.

An *increasing sequence* $(a_n)$ is one in which $a_{n+1} \geq a_n$ for all $n \in \mathbb{N}$. Likewise, a *decreasing sequence* $(a_n)$ is one in which $a_{n+1} \leq a_n$ for all $n \in \mathbb{N}$. The term *monotone* is used for a sequence which is either increasing or decreasing. In what follows it will, in fact, suffice to have a sequence which is 'eventually' monotone. In other words, the monotone behaviour occurs from some point of the sequence onwards. In any case, the convergence and associated limit of a given sequence is totally unaffected by the first few terms; or the first googolplex terms, for that matter! Finally, our deluge of new terms finishes with the notion of a *bounded* sequence—namely, a sequence $(a_n)$ in which $|a_n| \leq M$ for some fixed $M \in \mathbb{R}$, for all $n \in \mathbb{N}$. We can now prove, with the aid of our old friend the completeness axiom, that:

**Theorem 10.1** A bounded monotone sequence converges.

*Proof* Let $S = \{a_n : n \in \mathbb{N}\}$, the collection of all the terms of the sequence $(a_n)$. Since $S$ is a bounded set, both above and below, we can invoke the completeness axiom of Chapter 8. Hence, $S$ has an infinum $L'$ and a supremum $L$. If $(a_n)$ is increasing, we now claim that $(a_n) \to L$. If, on the other hand, $(a_n)$ is decreasing, then we claim that $(a_n) \to L'$. We prove the former and leave the details of the latter to the reader. So we have $L = \sup S$, and $a_{n+1} \geq a_n$ for all $n \in \mathbb{N}$. Now, $a_n \leq L$ for all $n \in \mathbb{N}$ and so, for any positive $\varepsilon$, $L - \varepsilon$ cannot be an upper bound for $S$ (why? [**10.8**]). Hence $\exists$ some $N = N(\varepsilon) \in \mathbb{N}$ with $a_N > L - \varepsilon$. Since $a_n$ is increasing, $n > N \Rightarrow a_n > L - \varepsilon$. So, for all $n > N$, $0 \leq L - a_n < \varepsilon$. Thus,

$$n > N \Rightarrow |a_n - L| < \varepsilon \qquad \square$$

So, just as the completeness axiom guaranteed the existence of irrational numbers, so it guarantees the existence of a limit for a bounded sequence.

## SEQUENCES AND SERIES

As an instructive example, we return to the sequence of example (2) of Section 10.1.

We have $a_1 = 3$ and $a_{n+1} = (a_n^2 + 5)/2a_n$ for $n \geq 1$. Hence,

$$a_{n+1} = (a_n^2 + (\sqrt{5})^2)/2a_n \geq 2\sqrt{5}a_n/2a_n = \sqrt{5}$$

using the inequality $(a^2 + b^2) \geq 2ab$. [10.9] Hence, $a_n \geq \sqrt{5}$ for all $n \in \mathbb{N}$. Moreover,

$$a_{n+1} \leq a_n \Leftrightarrow (a_n^2 + 5)/2a_n \leq a_n$$
$$\Leftrightarrow 5 \leq a_n^2$$

This latter statement is, of course, true. Thus, $(a_n)$ is a monotone decreasing sequence which is bounded below by zero. Hence, $(a_n)$ converges to some limit $L$. To find the limit, we need to resort to some of the algebraic manipulations mentioned before. First, since $(a_n) \to L$, then the sequence $(a_{n+1})$, which is just $(a_n)$ with the first term removed, must also converge to $L$. But $(a_{n+1}) = (a_n^2 + 5)/2a_n \to (L^2 + 5)/2L$ and so $L = (L^2 + 5)/2L$, leading to $L = \sqrt{5}$.

Incidentally, a similar argument to the one above shows that if the sequence of example (3) converged, then its limit would also be $\sqrt{5}$. But that sequence does not converge, and so we must always be sure that a given sequence has a limit before we begin any fancy algebraic manipulations to derive a fallacious answer. So much for our first test for convergence which handles monotone sequences. To cope with non-monotone sequences, we require the new notion of a *Cauchy sequence* $(x_n)$. This is one satisfying the following criterion:

---

Given any $\varepsilon > 0$, $\exists N = N(\varepsilon) \in \mathbb{N}$ such that $n, m \geq \mathbb{N} \Rightarrow |x_n - x_m| < \varepsilon$.

---

The intuitive interpretation of a Cauchy sequence is that, as we proceed, the terms get closer together. Then, surely, the $a_n$'s must all be getting closer to some fixed limit point? This is proved below, in our next theorem. The second half of the proof is optional reading, although the summary schedule is worth pondering.

**Theorem 10.2** A sequence $(x_n)$ converges $\Leftrightarrow (x_n)$ is a Cauchy sequence.

*Proof* To begin with, if we are given a convergent sequence $(x_n)$, then, given any $\varepsilon > 0$, $\exists N = N(\varepsilon/2)$ with $|x_n - L| < \varepsilon/2$ for all $n > N$. Hence,

$$|x_n - x_m| = |(x_n - L) - (x_m - L)|$$
$$\leq |x_n - L| + |x_m - L|$$
$$< \varepsilon/2 + \varepsilon/2 = \varepsilon \qquad \text{for all } n, m > N$$

Hence, every convergent sequence $(x_n)$ is a Cauchy sequence.

Our schedule for the reverse implication is to

(a) show that a Cauchy sequence is bounded;
(b) trap a subsequence of $(x_n)$ between an increasing and a decreasing sequence;
(c) invoke our result on bounded monotone sequences to show that the subsequence in (b) converges; and
(d) prove that the limit of the subsequence must be the limit of the (now convergent) Cauchy sequence.

Step (a): Put $\varepsilon = 1$ in the definition of a Cauchy sequence to deduce that $\exists N = N(1)$ such that $|x_n - x_m| < 1$ for all $n, m > N$. In particular, $|x_n - x_{N+1}| < 1$ for all $n > N$. Hence, each $x_n$ lies between $x_{N+1} - 1$ and $x_{N+1} + 1$. In other words, $(x_n)$ is bounded.

Steps (b) and (c): All the $x_n$'s now lie in some real interval $[a, b]$. Thus, an infinite number of the $x_n$'s lie in at least one of the right-hand and left-hand halves of $[a, b]$. Let $[a_1, b_1]$ denote a half which contains infinitely many $x_n$'s. Continue in this way (reminiscent of our approach to defining real numbers by closed rational intervals in Chapter 7), to obtain by bisection a nested sequence of real intervals $[a_1, b_1], [a_2, b_2], \ldots, [a_n, b_n], \ldots$, as in Figure 10.3.

By construction $(a_n)$ is a monotone increasing sequence which is bounded and $(b_n)$ is monotone decreasing and bounded. So, by our previous result, $(a_n) \to L$ and $(b_n) \to L'$. But

$$b_n - a_n = 2^{-1}(b_{n-1} - a_{n-1}) = \ldots = 2^{-n}(b - a)$$

and, hence, $(b_n - a_n) \to 0$. Thus, $L = L'$.

Our construction has also trapped an infinite number of the $x_n$'s in each interval of our nest and, hence, some subsequence of $(x_n)$ is forced to be convergent with limit $L$. To be precise, choose $y_1 = x_{n_1} \in [a_1, b_1]$. Since $[a_2, b_2]$ contains infinitely many $x_n$'s, we can choose $y_2 = x_{n_2} \in [a_2, b_2]$ with $n_2 > n_1$. Continuing in this way produces the required subsequence $(y_n)$.

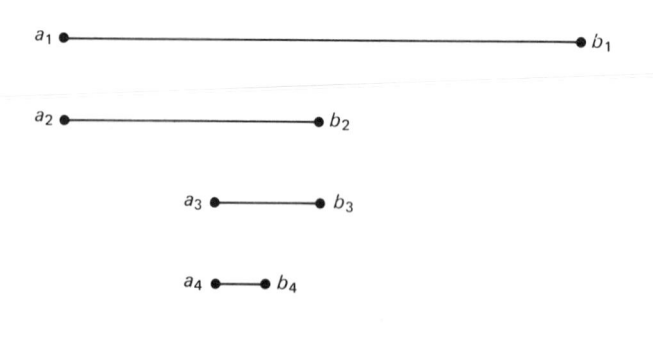

**Fig. 10.3**

Since $a_n \leq y_n \leq b_n$ for all $n \in \mathbb{N}$ and $(a_n) \to L$ and $(b_n) \to L$, a relatively straightforward $\varepsilon$-argument shows that $(y_n) \to L$. [**10.10**]

Step (d): Consider

$$|x_n - L| = |x_n - y_n + y_n - L|$$
$$\leq |x_n - y_n| + |y_n - L|$$

Since $(y_n) \to L$, for any given $\varepsilon > 0$, $\exists N_1 \in \mathbb{N}$ such that $|y_n - L| < \varepsilon/2$ for all $n > N_1$. Since $y_n = x_m$ for some $m$, then $|x_n - y_n| = |x_n - x_m| < \varepsilon/2$ for all $n, m > N_2$ for some $N_2 \in \mathbb{N}$. Hence $|x_n - L| < \varepsilon$ for all $n > N$ for some $N \in \mathbb{N}$. (In fact, any $N$ bigger than $N_1$ and $N_2$ will do.) Hence, $(x_n)$ is convergent, as required. □

In the above proof, we used at several points an inequality called the *triangle inequality* which is much used in $\varepsilon$-arguments. This inequality is that, for all real numbers $a$ and $b$, $|a \pm b| \leq |a| + |b|$. Try proving this by considering the squares of both sides. [**10.11**]

The brave reader who plodded through the preceding proof is to be congratulated. But be warned: analysis abounds with such proofs. We shall use this notion of Cauchy sequences to establish the convergence of the sequence derived from the Fibonacci sequence that we saw in example (4). Recall that $b_n = a_{n+1}/a_n$, where $a_1 = a_2 = 1$ and $a_{n+2} = a_{n+1} + a_n$, $n \geq 1$. Now,

$$b_{n+1} = a_{n+2}/a_{n+1} = (a_{n+1} + a_n)/a_{n+1} = 1 + a_n/a_{n+1} = 1 + 1/b_n$$

and, of course, $b_1 = 1$. First, we claim that $1 \leq b_n \leq 2$ for all $n \in \mathbb{N}$. Certainly this is true for $n = 1$, and if $1 \leq b_k \leq 2$ for some $k \in \mathbb{N}$, then $0.5 \leq 1/b_k \leq 1$ and so $1.5 \leq 1 + 1/b_k \leq 2$. In other words, $1.5 \leq b_{k+1} \leq 2$ also. Hence, by induction on $n$, $b_n$ lies between 1 and 2 for all $n \in \mathbb{N}$. Now,

$$|b_{n+1} - b_n| = |1 + 1/b_n - b_n| = |1 + (1 + 1/b_{n-1})^{-1} - (1 + 1/b_{n-1})|$$

So
$$|b_{n+1} - b_n| = |(b_{n-1}^2 - b_{n-1} - 1)/b_{n-1}(1 + b_{n-1})| \quad (1)$$

Also
$$|b_n - b_{n-1}| = |1 + 1/b_{n-1} - b_{n-1}| = |(b_{n-1}^2 - b_{n-1} - 1)/b_{n-1}| \quad (2)$$

From (1) and (2) $|(b_{n+1} - b_n)/(b_n - b_{n-1})| = |1/(1 + b_{n-1})| \leq 1/2$, since $b_{n-1} \geq 1$. Thus,

$$|b_{n+1} - b_n| \leq (1/2)|b_n - b_{n-1}|$$
$$\leq (1/2)^2 |b_{n-1} - b_{n-2}|$$
$$\leq$$
$$\vdots$$
$$\leq (1/2)^{n-1} |b_2 - b_1| = 2^{-n+1}$$

Now, by the triangle inequality,

$$|b_n - b_m| = |b_n - b_{n-1} + b_{n-1} - b_{n-2} + \cdots + b_{m+1} - b_m|$$
$$\leqslant 2^{-(n-2)} + 2^{-(n-3)} + \cdots + 2^{-(m-1)}$$
$$= 2^{-(m-1)}(1 + 1/2 + (1/2)^2 + \cdots + (1/2)^{n-m-1})$$
$$< 2^{-(m-1)}(1 - 1/2)^{-1} = 2^{-(m-2)}$$

Given $\varepsilon > 0$, choose $N = N(\varepsilon)$ so that $2^{-(N-2)} < \varepsilon$. So, for $n, m > N$, $|b_n - b_m| < \varepsilon$.

Hence, $(b_n)$ is indeed a Cauchy sequence and so it converges to some limit $L$. After all this excitement, it may seem a bit of an anticlimax to actually calculate the limit. We use a previous trick. Since $(b_n) \to L$, then $(b_{n+1}) \to L$ also. But $(b_{n+1}) = (1 + 1/b_n)$ also converges to $1 + L^{-1}$, provided that $L \neq 0$. Why is $L \neq 0$? [**10.12**] Thus, $L = 1 + L^{-1}$, leading to $L = \frac{1}{2}(1 + \sqrt{5})$, an irrational number. This number is called the *golden ratio* and was known to the ancient Greeks. It occurs repeatedly in many geometrical situations.

> 'Geometry has two great treasures: one is the theorem of Pythagoras;
> the other, the division of a line into extreme and mean ratio.'
>
> Kepler (1571–1630)

Referring to Figure 10.4, P is the point which divides AB such that $AB/AP = AP/PB$, the golden ratio. So $(a+b)/a = a/b = L$, say. This leads to $L = 1 + L^{-1}$ and so $L = \frac{1}{2}(1 + \sqrt{5})$. A partial anthology of its occurrences now follows.

Any two diagonals of a regular pentagon divide each other in the golden ratio. The golden rectangle (i.e. one whose sides are in the golden ratio) is regarded as being the most aesthetically pleasing and it appears more often in art and architecture than any other rectangle. There is ample evidence that the Greeks incorporated the golden rectangle into their artefacts, as the Parthenon at Athens demonstrates. We derived the golden ratio by looking at successive terms of the Fibonacci sequence, itself arising from a problem concerning the breeding of rabbits. We should not, therefore, be surprised if Nature herself, with her many varied growth processes, provides us with more examples of the golden ratio. The number of opposing spirals found on a sunflower or a pine cone are consecutive Fibonacci numbers. The spirals on a snail's shell are such that each quarter-revolution is contained in a square removed from a golden rectangle. All the rectangles in Figure 10.5 are golden. For more details see *The Divine Proportion*, by H. E. Huntley.

The golden ratio is an irrational number, and the ratios between successive terms of the Fibonacci sequence are rationals. Our work on Cauchy sequences,

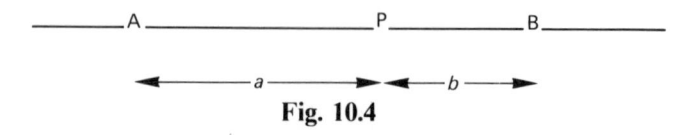

Fig. 10.4

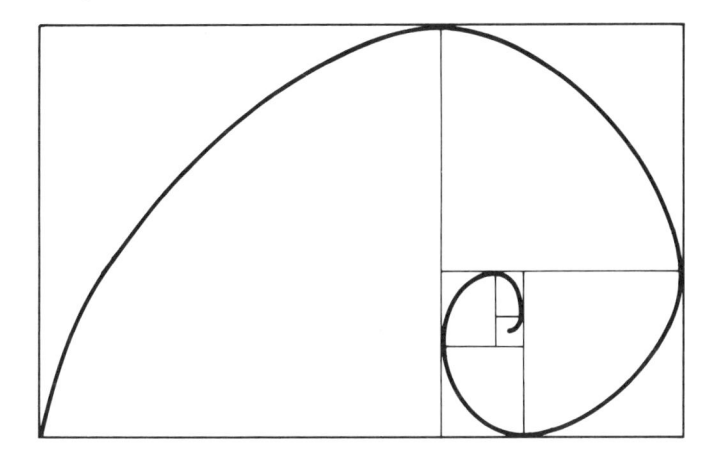

**Fig. 10.5** An equiangular spiral

involving use of the completeness axiom, has enabled us to show that the golden ratio is the limiting value of consecutive ratios of the Fibonacci numbers. The completeness axiom is, of course, the one which guarantees the existence of the irrational golden number. In our final section and the next chapter, we shall see again and again that the completeness axiom really is the keystone in the vast edifice of real analysis.

## 10.3 SERIES

Consider the following arguments

(a) If
$$S = 1 + 1/2 + 1/4 + 1/8 + \cdots + 1/2^n + \cdots$$
then
$$S/2 = 1/2 + 1/4 + \cdots + 1/2^{n+1} + \cdots$$
Subtraction yields $S/2 = 1$, and so $S = 2$.

(b) If
$$S = 1 - 1 + 1 - 1 + \cdots + (-1)^n + \cdots$$
then
$$S = (1-1) + (1-1) + \cdots$$
Hence, $S = 0$. But
$$S = 1 + (-1+1) + (-1+1) + \cdots$$
and so $S = 1$.

151

Even worse, by the binomial expansion,
$$(1-x)^{-1} = 1 + x + x^2 + \cdots$$
In particular, when $x = -1$,
$$1/2 = 1 - 1 + 1 - 1 + \cdots$$
In other words, $S = 1/2$.

(c) Using the binomial expansion above, when $x = 2$,
$$-1 = 1 + 2 + 4 + 8 + \cdots$$

(d) Once more we exploit the poor overworked binomial expansion:
$$x/(1-x) = x(1-x)^{-1} = x + x^2 + x^3 + \cdots$$
$$x/(x-1) = (1 - 1/x)^{-1} = 1 + 1/x + 1/x^2 + \cdots$$
Adding gives
$$0 = \cdots + 1/x^2 + 1/x + 1 + x + x^2 + \cdots$$

What is going on? Only argument (a) is, in some sense, correct. We have been doing ordinary algebra with infinite sums without even considering what an infinite sum might be. To be explicit, in (a) we extended the distributive law $a \cdot (b + c) = a \cdot b + a \cdot c$ to hold when we wished to multiply an infinite sum by a fixed number. Can this be done? In (b) we rebracketed an infinite sum in different ways and obtained different answers. The use of the binomial expansion to obtain the answer $1/2$ was a fact that puzzled Euler. The observation that $1/2$ is the average of 0 and 1 is clearly no way out of our dilemma! Results (c) and (d) also caused puzzlement to Euler and his eighteenth-century colleagues. Indeed, seeming paradoxes such as these led directly into the nineteenth-century effort to inject rigour into mathematics. We in the twentieth century suffer (or enjoy, depending on your point of view) the consequences of this crisis in mathematics. Many of the paradoxes can be resolved by a proper formulation of what we mean by *an infinite sum*. Even when such sums do have a meaning, care will need to be exercised when using them. The first barrier to an understanding of when an infinite sum is meaningful or not is perhaps the notation. Given an infinite sequence of real numbers $a_1, a_2, a_3, \ldots$, the infinite series, $\sum_{k=1}^{\infty} a_k$, is meant to denote the 'sum' of those numbers. Clearly, this instruction is impossible to implement. In fact, the more and more terms of the original sequence we add together, the larger and larger becomes the 'sum-so-far'. In many instances the sum of sufficiently many of the $a_i$'s continues to grow out of control and so $\sum_{k=1}^{\infty} a_k$ does not exist! So what does $\sum_{k=1}^{\infty} a_k$ actually mean?

Let $s_1 = a_1$, $s_2 = a_1 + a_2$, $s_3 = a_1 + a_2 + a_3$, $\cdots$ and, in general, $s_n = a_1 + a_2 + \cdots + a_n$. $s_n$ is called a *partial sum*, a perfectly well-defined finite sum. If, and it is a big if, the sequence $(s_n)$ of $n$th partial sums converges (as a sequence), then we say that the infinite series $\sum_{k=1}^{\infty} a_k$ is a *convergent series*. Moreover, we call $S = \lim_{n \to \infty}(s_n)$ the *sum* of that series. So, the sum of an infinite series is to be interpreted as the limit of its partial sums. If those partial sums constitute a divergent sequence (by which we mean non-convergent sequence), then the associated series is called a divergent series. In this case we say, rather paradoxically, that $\sum_{k=1}^{\infty} a_k$ does not exist!

As our first example we re-examine (a) and put the whole argument on a sound footing. So consider $\sum_{k=0}^{\infty} (1/2)^k$ and don't worry about beginning the count at 0; all the best mathematicians do! The $n$th partial sum is given by

$$s_n = 1 + (1/2) + (1/2)^2 + \cdots + (1/2)^{n-1}$$

So

$$\tfrac{1}{2}s_n = (1/2) + (1/2)^2 + \cdots + (1/2)^n$$

and subtraction gives, after rearrangement, $s_n = 2(1 - (1/2)^n)$. Now, $(s_n) \to 2$ as $n \to \infty$ and so $\sum_{k=0}^{\infty} (1/2)^k$ is a convergent series with sum 2. The intuitive interpretation of our last statement is that the more and more terms of the series that we sum, the closer and closer our answer gets to 2. However many terms that we actually sum, we can never achieve the answer 2. This example of seeming to add up a larger and larger number of smaller and smaller bits, yet never exceeding a fixed finite quantity, is the basis of many paradoxes. The famous Greek paradox, posed by Zeno, of the race between Achilles and a tortoise is an example. This is not dissimilar to the terrible trio's dilemma in Section 8.5.

So we have our first convergent series, $\sum_{k=0}^{\infty}(1/2)^k = 2$. This is a special case of the *geometric series* $\sum_{k=0}^{\infty} x^k$. Investigation of its partial sums reveals that convergence only occurs when $|x| < 1$. The sum for such $x$ is $(1-x)^{-1}$. The proof proceeds along lines similar to ours for the case $x = 1/2$. [**10.13**] You should now be in a position to resolve the paradoxes in (c) and the last part of (b). [**10.14**] The only unexplained behaviour in argument (b) is concerned with the series $\sum_{k=0}^{\infty} (-1)^k = 1 - 1 + 1 - 1 + \cdots$. The $n$th partial sums here are $s_n = 0$, if $n$ is even, and $s_n = 1$, if $n$ is odd. Thus, $(s_n)$ is for us a divergent sequence and so the series diverges and so can have no sum. Rebracketing of the terms in different ways amounted to restricting our attention to either the even or the odd partial sums only and thus leading us to a false conclusion. Looked at another way, in (b) we were presented with three infinite series, $\sum_{k=0}^{\infty}(1-1)$, $1 + \sum_{k=0}^{\infty}(-1+1)$ and $\sum_{k=0}^{\infty}(-1)^k$. The first two converge with sums 0 and 1, respectively, and the third is divergent. We were tricked into assuming that all three series were the same. The following example is in the same vein.

**Example** The series $\sum_{k=1}^{\infty} 1/(2k-1)(2k+1)$ is convergent. In fact, since

$$1/(2k-1)(2k+1) = 1/2(2k-1) - 1/2(2k+1)$$

$$s_n = \tfrac{1}{2}((1/1 - 1/3) + (1/3 - 1/5) + \cdots + (1/(2n-1) - 1/(2n+1)))$$
$$= \tfrac{1}{2}(1 - (1/(2n+1))) \to \tfrac{1}{2} \quad \text{as} \quad n \to \infty$$

Alternatively, using

$$1/(2k-1)(2k+1) = k/(2k-1) - (n+1)/(2n+1)))$$
$$s_n = ((1/1 - 2/3) + (2/3 - 3/5) + \cdots + (n/(2n-1) - (n+1)/(2n+1)))$$
$$= 1 - (n+1)/(2n+1) \to \tfrac{1}{2} \quad \text{as} \quad n \to \infty$$

However, if we are sloppy and write $S$ for the sum of the series and proceed as follows, we can be misled:

$$S = 1/1 \cdot 3 + 1/3 \cdot 5 + 1/5 \cdot 7 + \cdots$$
$$= \tfrac{1}{2}((1/1 - 1/3) + (1/3 - 1/5) + \cdots$$
$$= \tfrac{1}{2}$$

But

$$S = 1/1 \cdot 3 + 1/3 \cdot 5 + 1/5 \cdot 7 + \cdots$$
$$= (1/1 - 2/3) + (2/3 - 3/5) + \cdots$$
$$= 1$$

Once again there are three different series hiding in this example, and we shall return to them after our discussion of absolute convergence. The moral is, perhaps, to just beware of the plus, dot, dot, dot, ....

The geometric series that we saw earlier is an example of what we call a *power series*. It converged to a function of $x$—namely, its sum, $(1-x)^{-1}$—for $x$ inside a particular interval of real numbers, and diverged for $x$ outside that interval. Many of the elementary functions in mathematics, $\sin x$, $e^x$, $(1+x)^t$, etc., can be expressed as the sums of suitable power series provided that the $x$'s involved are suitably restricted. For example, it can be shown that $\sum_{k=1}^{\infty} (-1)^{k+1}(x^k/k)$ converges if and only if $-1 < x \leq 1$ and that for such $x$ its sum is $\log_e(1+x)$.

Naturally, when using such expressions, we must respect the restrictions on $x$. In paradox (d) our calculations are valid but, unfortunately, the set of $x$'s for which the conclusion holds is the empty set!

Much of the theory of convergent sequences can be carried over to give us tests for the convergence of infinite series, and a seemingly unending list of such tests can be found in the literature on analysis. Even when we have such tests and have amassed a clutch of known convergent series, we must still proceed with caution and not blindly apply the ordinary rules of algebra (or calculus, for that matter). In particular, the temptation to rearrange the terms of an infinite series, to more easily facilitate its summation, is strong. But consider the following argument.

Given that $\log_e 2 = 1 - 1/2 + 1/3 - 1/4 + \cdots$, or, more precisely, that $\sum_{k=1}^{\infty} (-1)^{k+1}/k$ converges to $\log_e 2$, we rearrange terms to obtain

$$\log_e 2 = (1 + 1/3 + 1/5 + \cdots) - (1/2 + 1/4 + 1/6 + \cdots)$$
$$= (1 + 1/3 + 1/5 + \cdots) + (1/2 + 1/4 + 1/6 + \cdots)$$
$$\quad - 2(1/2 + 1/4 + 1/6 + \cdots)$$
$$= (1 + 1/2 + 1/3 + 1/4 + \cdots) - (1 + 1/2 + 1/3 + 1/4 + \cdots)$$
$$= 0$$

So $\log_e 2 = 0$, a clear absurdity!

This situation is far worse than our single example suggests. A rearrangement of a given series $\sum_{k=1}^{\infty} a_k$ is another series, containing exactly the same terms as the original series, but in a different order. Such a rearrangement can result in a series whose sum is anything we like. It can even result in a divergent series. It appears that rearrangement should be banned, which is a pity, since algebraic manipulations are usually needed in simplifying many finite problems. Luckily, there is a large class of series whose terms can be rearranged, without affecting either the convergence or the sum. Such series are called *absolutely convergent* and the condition that guarantees all that we have claimed for such series is as follows:

$$\sum_{k=1}^{\infty} a_k \text{ is absolutely convergent} \Leftrightarrow \sum_{k=1}^{\infty} |a_k| \text{ is convergent}$$

Absolutely convergent series (all convergent series with positive terms must rather trivially be absolutely convergent) are necessarily convergent, but not conversely. A series which is convergent but not absolutely convergent is called *conditionally convergent*. Such a series must contain a liberal sprinkling of both + and − signs among its terms. For example,

$$\sum_{k=1}^{\infty} (-1)^{k+1}/k = 1 - 1/2 + 1/3 - 1/4 + \cdots$$

is convergent, as discussed earlier, but

$$\sum_{k=1}^{\infty} \frac{1}{k} = 1 + \frac{1}{2} + \frac{1}{3} + \frac{1}{4} + \cdots$$

is, as we shall see, a divergent series. Hence, $\sum_{k=1}^{\infty}(-1)^{k+1}/k$ is conditionally convergent and so rearrangement of its terms is banned!

On the other hand, $\sum_{k=1}^{\infty}(-1)^{k+1}/k^2 = 1 - 1/4 + 1/9 - 1/16 + \cdots$ is also convergent. Its sum is $\pi^2/12$, a result requiring more advanced calculus. The associated series of absolute values, $\sum_{k=1}^{\infty} 1/k^2$ turns out to be convergent (its sum is $\pi^2/6$) and so $\sum_{k=1}^{\infty}(-1)^{k+1}/k^2$ is absolutely convergent.

Suppose, then, that we have a series $\sum_{k=1}^{\infty} a_k$ which contains an infinite number of both positive and negative terms. Denote formally by $S^+$ and $S^-$ the infinite series formed by summing (or attempting to sum) the positive and negative terms, respectively. If $\sum_{k=1}^{\infty} a_k$ is absolutely convergent, then the positive and negative pieces also converge [**10.15**] and their sums are $S^+$ and $S^-$, respectively. For a conditionally convergent series, either one or both of these positive and negative pieces must diverge. [**10.16**] As an illustration, consider the following three series:

(1) $\frac{1}{2}(1/1 - 1/3 + 1/3 - 1/5 \cdots)$

(2) $1/1 - 2/3 + 2/3 - 3/5 + \cdots$

(3) $1/1 \cdot 3 + 1/3 \cdot 5 + 1/5 \cdot 7 + \cdots$

They are all different series which, by rebracketing, we were previously led to believe were all the same. Series (3) is convergent, as we have seen, and since all its terms are positive, it must be absolutely convergent. Series (1) also turns out to be absolutely convergent. Series (2) is divergent. Quite accidentally, series (1) and (3) both have sum $1/2$.

To round off this chapter, we shall fill in some of the holes we have left. We establish the convergence of a class of series some of which we have seen. Finding their actual sums is a task we shall not embark on. In general, manipulation of partial sums to prove convergence and find the sum of the series will only work if we can express the partial sums in a nice closed form. That's how we summed the geometric series. One would be hard put to it to obtain by a similar method some of the sums we have quoted: $\log_e 2$, $\pi^2/6$, and so on. This is because the sums of series arise the other way around. That is to say, some function of $x$ has been expanded as, say, a power series in $x$ which converges (for certain $x$) to the original function. These are the so-called *Taylor–Maclaurin series*. The one for $\log_e(1 + x)$ was quoted earlier in this chapter. Quite often, for different purposes, functions are realised as the sums of infinite series involving, not powers of $x$, but other functions of $x$ such as sine and cosine. Again these *Fourier series* are valid for certain $x$ only, and can be used to deliver up the sums of certain infinite series. A concrete example of a Fourier series is

$$\frac{\pi^2}{3} + \sum_{k=1}^{\infty} (4(-1)^k \cos kx)/k^2$$

which converges (absolutely) to $x^2$ when $-\pi \leqslant x \leqslant \pi$. Putting $x$ equal to $\pi$ and 0, respectively, yields the results

$$\sum_{k=1}^{\infty} \frac{(-1)^{k+1}}{k^2} = \frac{\pi^2}{12} \quad \text{and} \quad \sum_{k=1}^{\infty} \frac{1}{k^2} = \frac{\pi^2}{6}$$

To date, we have used the facts that

$$\sum_{k=1}^{\infty} \frac{1}{k} = 1 + \frac{1}{2} + \frac{1}{3} + \frac{1}{4} + \cdots$$

is divergent, while

$$\sum_{k=1}^{\infty} \frac{1}{k^2} = 1 + \frac{1}{4} + \frac{1}{9} + \frac{1}{16} + \cdots$$

is convergent. The latter is believable—adding on smaller and smaller bits makes relatively little difference to the sum of the first few terms: the former does not strike one as being at all obvious! $\sum_{k=1}^{\infty} 1/k$ is called the *harmonic series* and its divergence is critical in the sense that the *p-series*, $\sum_{k=1}^{\infty} 1/k^p$, converge if and only if $p > 1$. We examine in detail the cases $p = 1$ and $p = 2$, and produce both geometric and algebraic arguments for the divergence of one and the convergence of the other.

## Argument 1 (Geometric)

Consider $f(x) = 1/x$ for $x > 0$, as in Figure 10.6. The rectangles indicated have areas $1, 1/2, 1/3, \ldots$, etc. Hence, the $n$th partial sum, $s_n$, of $\sum_{k=1}^{\infty} 1/k$ is just the sum of the areas of the first $n$ rectangles. But then

$$s_n \geq \int_1^{n+1} \frac{1}{x} dx = [\log_e x]_1^{n+1} = \log_e(n+1)$$

In other words, $(s_n)$ diverges, since $\log_e$ is a strictly increasing and unbounded function.

Now look at $f(x) = 1/x^2$ for $x > 0$, as in Figure 10.7. The rectangles indicated have areas $1/2^2, 1/3^2, \ldots$, etc. Hence, the $n$th partial sum, $s_n$, of $\sum_{k=1}^{\infty} 1/k^2$ is 1 plus the sum of the areas of the first $(n-1)$ rectangles. Hence,

$$s_n \leq 1 + \int_1^n 1/x^2 \, dx = 1 + [-1/x]_1^n = 2 - 1/n$$

for all $n \in \mathbb{N}$. In other words, $(s_n)$ is a monotone increasing (why? [**10.17**]) sequence bounded above by 2. Hence, $(s_n) \to L$ for some $L$, $0 < L < 2$. In fact, $L = \pi^2/6 = 1.64$ approximately, as we quoted earlier.

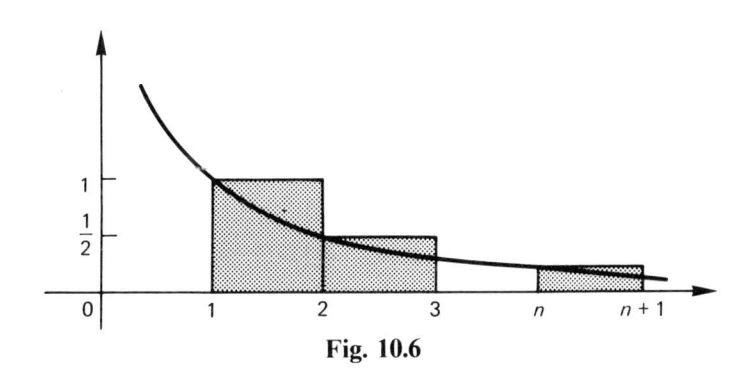

**Fig. 10.6**

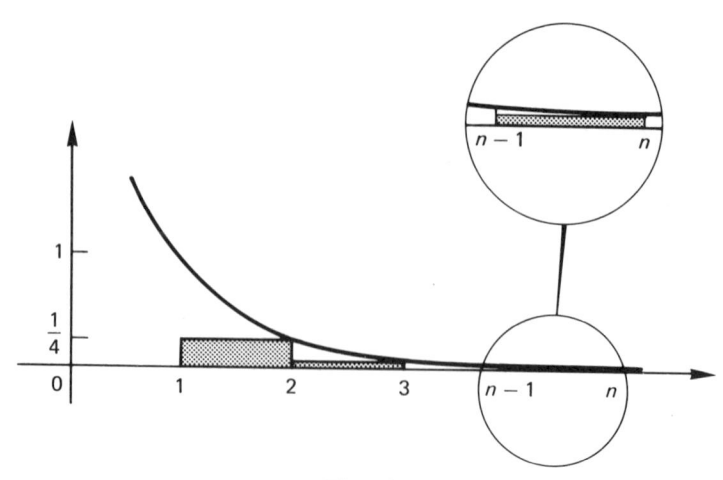

Fig. 10.7

## Argument 2 (Algebraic)

The $2^n$th partial sum of $\sum_{k=1}^{\infty} 1/k$ is

$$s_{2^n} = 1 + 1/2 + \cdots + 1/2^n$$
$$= 1 + 1/2 + (1/3 + 1/4) + (1/5 + \cdots + 1/8) + \cdots$$
$$\quad + (1/(2^{n-1} + 1) + \cdots + 1/2^n)$$
$$\geq 1 + 1/2 + (1/4 + 1/4) + (1/8 + \cdots + 1/8) + \cdots$$
$$\quad + (1/2^n + \cdots + 1/2^n)$$
$$= 1 + 1/2 + 2/4 + 4/8 + \cdots + 2^{n-1}/2^n$$
$$= 1 + n/2$$

The partial sums are unbounded, and, hence, $\sum_{k=1}^{\infty} 1/k$ diverges. For example, we can estimate the number of terms that need to be summed in order to exceed any large number. So, $s_{2^{100}} \geq 51$, $s_{2^{1000000}} \geq 500\,001\ldots$ visions of googolplexes.

By contrast, the $s_{2^n-1}$ partial sum of $\sum_{k=1}^{\infty} 1/k^2$ is

$$s_{2^n-1} = 1 + 1/2^2 + \cdots + 1/(2^n - 1)^2$$
$$= 1 + (1/2^2 + 1/3^2) + (1/4^2 + \cdots + 1/7^2) + \cdots$$
$$\quad + (1/2^{2(n-1)} + \cdots + 1/(2^n - 1)^2)$$
$$\leq 1 + (1/2^2 + 1/2^2) + (1/4^2 + \cdots + 1/4^2) + \cdots$$
$$\quad + (1/2^{2n-2} + \cdots + 1/2^{2n-2})$$
$$= 1 + 2/2^2 + 4/4^2 + \cdots + 2^{n-1}/2^{2n-2}$$

$$= 1 + 1/2 + (1/2)^2 + \cdots + (1/2)^{n-1}$$
$$= 2(1 - (1/2)^n) \leqslant 2$$

So the partial sums of $\sum_{k=1}^{\infty} 1/k^2$ are bounded and convergence is thus guaranteed.

These algebraic arguments are perhaps more rigorous than the geometric ones but both are valid. The former is an application of integration often used to investigate infinite sums. A rather more entertaining proof that $\sum_{k=1}^{\infty} 1/k$ diverges is our final argument.

## Argument 3 (Fun)

Consider

$$\tfrac{16}{11} = 1 + (\tfrac{16}{11} - 1) = 1 + \tfrac{5}{11}$$
$$\tfrac{5}{11} = \tfrac{1}{3} + (\tfrac{5}{11} - \tfrac{1}{3}) = \tfrac{1}{3} + \tfrac{4}{33}$$
$$\tfrac{4}{33} = \tfrac{1}{9} + (\tfrac{4}{33} - \tfrac{1}{9}) = \tfrac{1}{9} + \tfrac{3}{297}$$
$$\tfrac{3}{297} = \tfrac{1}{99}$$

Hence,

$$\tfrac{16}{11} = 1 + \tfrac{1}{3} + \tfrac{1}{9} + \tfrac{1}{99}$$

It is a remarkable fact that every rational can be expressed as a sum of a finite number of distinct rationals of the form $1/n$. At each stage of our example, except the last, we had a fraction present of the form $p/q$, $p \neq 1$. We then expressed $p/q$ as the sum of the next largest rational member of the set $\{1, 1/2, 1/3, 1/4, \ldots\}$ which was less than $p/q$ and another rational $r/s$. This process terminated after a finite number of steps. So, given any rational $R$, there exists an $N \in \mathbb{N}$ such that $R$ is the sum of some of the fractions of the form $1/n$ up to and including $1/N$. But then $s_N \geqslant R$, where $s_N$ is a partial sum of $\sum_{n=1}^{\infty} 1/k$, and so the harmonic series is divergent, since its partial sums are unbounded.

Returning briefly to our first proof that $\sum_{k=1}^{\infty} 1/k$ is divergent, we can find an interesting clue to the growth rate of the $n$th partial sums. Recall that the partial sums were related to the area under the graph of $f(x) = 1/x$. The integral of $f(x)$ is, of course, $\log_e x$. So perhaps the rate of increase of $(1 + 1/2 + 1/3 + 1/4 + \cdots + 1/n)$ is approximately the rate of increase of $\log_e n$?

Appealing to Figure 10.6, we can deduce that

$$1 + 1/2 + 1/3 + 1/4 + \cdots + 1/n \geqslant \int_1^{n+1} \frac{1}{x} dx = \log_e(n+1)$$

for all $n \in \mathbb{N}$.

Now let $\beta_n = (1 + 1/2 + 1/3 + 1/4 + \cdots + 1/n) - \log_e n$. Immediately the above inequality gives that $\beta_n \geq \int_n^{n+1} \frac{1}{x} dx > 0$ for all $n \in \mathbb{N}$.

Also,

$$\beta_{n+1} - \beta_n = 1/(n+1) - \log_e(n+1) + \log_e n$$

$$= 1/(n+1) - \int_n^{n+1} \frac{1}{x} dx < 0$$

by looking at the last rectangle on Figure 10.6. Hence, $(\beta_n)$ is decreasing and bounded below. Hence, $(\beta_n) \to \beta$ as $n \to \infty$. In fact, $\beta$ is approximately equal to 0.5771 and is known as *Euler's number*. It is as yet unknown whether $\beta$ is irrational; on the other hand, each $\beta_n$ is irrational.

We hope that in this chapter we have revealed the wonders of the infinitesimal. We have merely scratched at the surface of the topic of convergence of sequences and series. However, we have sowed the seeds of some important concepts and fine-tuned your intuition.

# 11 Graphs and Continuity—in which we arrange a marriage between Intuition and Rigour

'It then only remained to discover...a real definition of the essence of continuity. I succeeded 24 November 1858, and a few days afterward I communicated the results to my dear friend Durege with whom I had a long and lively discussion.'

Richard Dedekind

## 11.1 INTRODUCTION

The mathematical idea of continuity is analogous to, but not the same as, the intuitive idea of continuity which we associate with time, space or motion. We think of time as unbroken, of space as smooth with no holes and of motion as uninterrupted. The mathematician, perverse as ever, seeks to redefine this comfortable vague notion of continuity by a more useful, more precise but more troublesome definition. The real line, which we have taken such pains to define, is deemed to be continuous. Recall that we required the completeness axiom to plug the imperceptible gaps. In this chapter we shall be mainly concerned with the notion of a 'continuous function'. The definition of this concept is necessarily precise but it accords, most of the time, with our notion of an unbroken curve. The current theory of mathematical continuity is an abstract logical edifice which may or may not describe the way space actually is. So far, mathematicians have been able to resolve any unexpected quirks of the rigorously defined concept of a continuous function more or less to everyone's satisfaction. One of the founders of analysis, a Catholic priest, Bernhard Bolzano (1781–1848), when analysing the paradoxes of the infinite, was driven to define various intuitively obvious mathematical concepts such as continuity. Typical of the sort of results he deduced is the following *intermediate value theorem*:

> A function $f$ which is continuous on some closed interval which takes both positive and negative values on that interval must be zero at least once in that interval.

This far from world-shattering result is tantamount to saying that between the basement and first floor of a building there must exist a ground floor. This is perhaps one of the most offputting features of modern analysis courses.

They proceed through the painful process of proving intuitively obvious properties of continuous functions from the chosen definition of mathematical continuity. There are many spinoffs; for example, one becomes more and more convinced that mathematical continuity does give a good description of situations or processes exhibiting intuitive continuity. Also, the elementary theorems needed to develop the concept towards applications are in some sense themselves key ideas which capture the very essence of continuity.

## 11.2 GRAPHS

Given a function $f$ specified by some formula $f(x)$, it is theoretically possible to display the values of $f$ on a two-dimensional picture, called the *graph* of $f$. In this chapter our function will have domain a subset of the real numbers and codomain the real numbers. In other words, for certain specified $x$'s in the set of real numbers, the formula $f(x)$ will specify uniquely determined real values. Hence, the graph of $f$, which represents $f$ itself, is the set

$$S = \{(x, f(x)) : x \text{ is in the domain of } f\}$$

Since the domain of $f$ will usually be an infinite set, it appears at first quite daunting to actually draw all these points in the Cartesian plane. As we shall see, for many functions $f$ the task is somewhat simpler than it appears, except that the old problem of 'Are there any gaps?' reappears. To be more precise, consider the following sequence of examples.

(1) Suppose that $f(x) = mx$, where $m$ is a fixed real constant and $x$ can take any real value. In Figure 11.1 we have plotted two points, O and A, which lie on the graph of $f$. We have also drawn an infinite straight line $L$ passing through O and A. Now consider any point A'$(a, b)$ lying on $L$. Since $\triangle$OA'B' is similar to $\triangle$OAB (they have the same angles), we see that $a/x = b/mx$. Hence, A' is the point $(a, ma)$ and so A' lies on the graph of $f$. The evidence is very strong; the graph of $f$ is the line $L$, which, geometrically

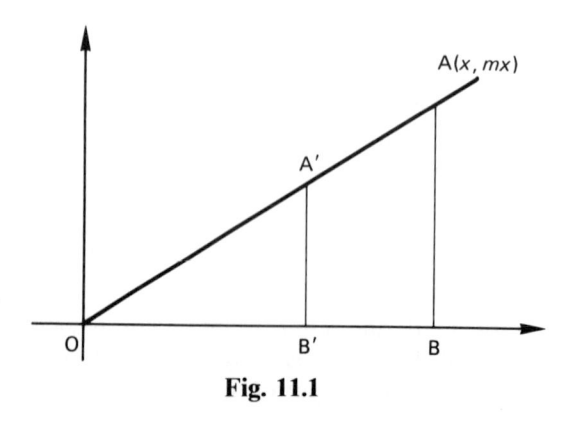

**Fig. 11.1**

at least, appears to be 'continuous'. We are not, of course, claiming that this latter fact has been proved. It would be possible for us to present a purely formal algebraic argument that the graph of $f$ is indeed a copy of $\mathbb{R}$ inclined at an angle $\tan^{-1} m$ to the $x$-axis. Then the completeness of $\mathbb{R}$, the existence of 'no gaps', can be taken to be synonymous with our claim that the graph of $f$ is a 'continuous' line.

(2) Consider the function $g(x) = x^2$, which squares any real number to produce a positive or zero real number. Some points on the graph of $g$ are plotted in Figure 11.2. Plotting more and more points suggests that the graph of $g$ is the continuous curve we call a *parabola*. We need a more convincing argument, as we had in example (1). First of all,

$$0 < x_1 < x_2 \Rightarrow 0 < x_1^2 < x_2^2$$

means that in the positive quadrant any point of the graph to the right of a previously plotted point will be higher. It also seems clear that if $x_2$ is close to $x_1$, then $x_2^2$ will be reasonably close to $x_1^2$. So no sudden jumps are expected. Once again we are using vague notions of closeness to justify our gut feeling that the graph of $g$ ought to be the continuous (and smooth) curve depicted in Figure 11.3.

(3) In this example $[x]$ will denote the greatest integer less than or equal to the real number $x$. For instance, $[2 \cdot 1] = 2$, $[-3 \cdot 1] = -4$. We consider the function $h$, given by $h(x) = x - [x]$. For $n \leq x < n+1$, where $n \in \mathbb{N}$, we can write $h(x) = x - n$. The graph of $h$ over this range is a straight line. The graph of $h$ is given in Figure 11.4. Notice the finite jumps at integer values of $x$. Any sensible definition of a continuous graph must highlight discontinuous jumps such as these. Have a go at graphing the following. [**11.1**]

$$h_1(x) = \sqrt{(x - [x])} \quad \text{and} \quad h_2(x) = [x] + \sqrt{(x - [x])}$$

The function $h$ in the previous example is a *periodic* function. That is to say, one whose graph over a certain range is then repeated *ad infinitum* across

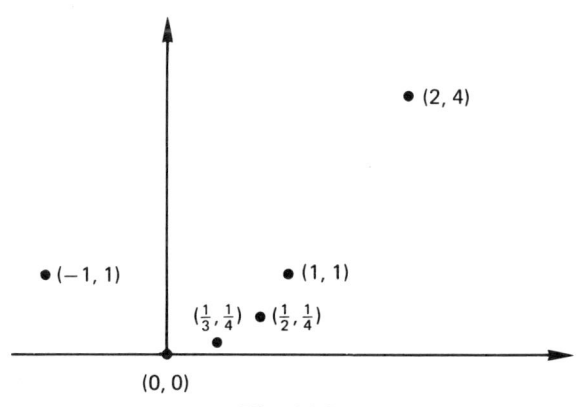

• (2, 4)

• (−1, 1)    • (1, 1)

($\frac{1}{3}, \frac{1}{4}$)  • ($\frac{1}{2}, \frac{1}{4}$)

(0, 0)

Fig. 11.2

**Fig. 11.3**

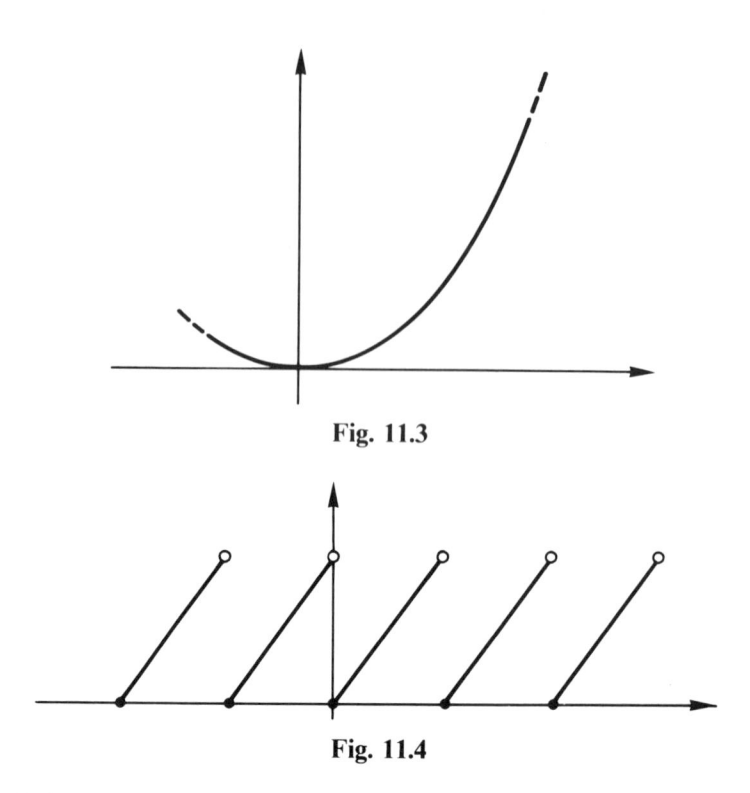

**Fig. 11.4**

the whole domain of $h$. Readers are perhaps more familiar with the periodic functions $\sin x$ and $\cos x$. The function $\sin x$ is usually defined as the vertical displacement of a point P on a wheel of unit radius as it rotates in a uniform continuous manner; $x$ being the angular displacement. See Figure 11.5. Clearly, after one revolution ($x$ having moved through $2\pi$ radians) the values of $\sin x$ repeat. Hence, $\sin x$ is periodic. Other symmetrical properties of the graph can be deduced from the motion of the wheel. More importantly, we are led to believe that the graph of $\sin x$ is a continuous curve because the

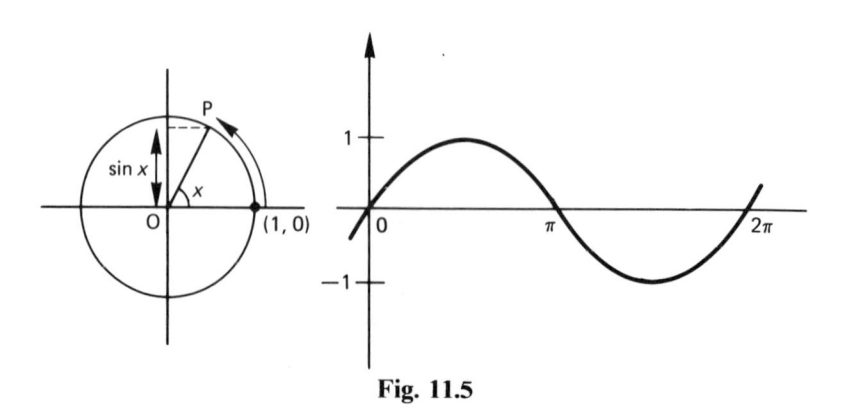

**Fig. 11.5**

motion of the wheel was continuous with respect to time and space. Our forthcoming definition of continuity will support these beliefs, and in fact all the elementary functions are continuous (except possibly at a finite number of values of $x$). It is precisely because these elementary functions are almost everywhere continuous and differentiable and many other things, that they have been given particular names. One could argue that in the real world such functions are the exception rather than the rule (see Section 11.4), and this is one very good reason why such a fuss should be made of formulating a satisfactory mathematical notion of continuity. If the elementary functions (which include polynomials, trigonometric functions, logarithmic and exponential functions) were all the ones we ever needed and they are seen to be continuous on intuitive grounds, then why bother with a rigorous mathematical definition of continuity? To return to our examples.

(4) Let $k(x) = \sin(1/x)$, where $x$ is a non-zero real number. We compute some values of $k$. First, $k(x) = 0$ whenever $1/x = n\pi$ for any non-zero integer $n$. Hence,

$$k(x) = 0 \quad \text{when} \quad x \in \{1/n\pi : n \in \mathbb{Z}, n \neq 0\}$$

Similarly,

$$k(x) = 1 \quad \text{when} \quad x \in \{2/(4n+1)\pi : n \in \mathbb{Z}\}$$

and

$$k(x) = -1 \quad \text{when} \quad x \in \{2/(4n-1)\pi : n \in \mathbb{Z}\}$$

The graph of $k$ oscillates between 1 and $-1$, and the frequency increases ever more rapidly as we approach $x = 0$, where $k$ is undefined. For $x \neq 0$, one can convince oneself that the graph of $k$ is as depicted in Figure 11.6.

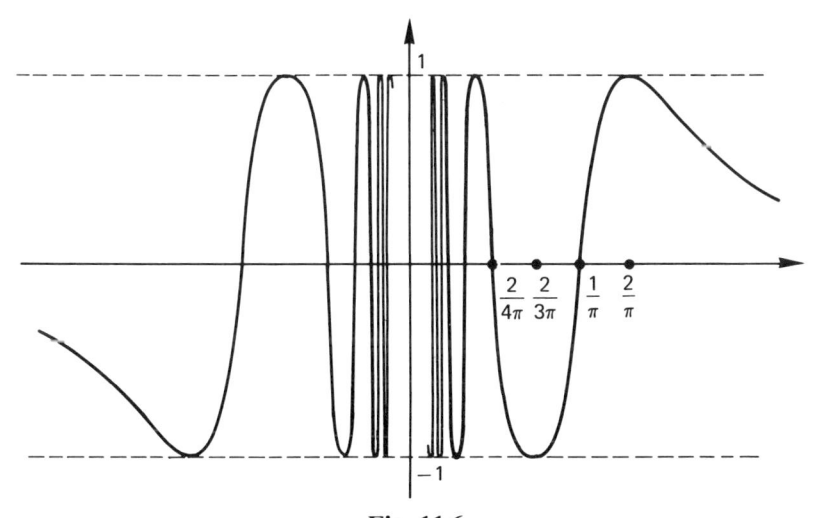

**Fig. 11.6**

But is $k$ continuous at $x = 0$? If we let $l(x) = x \cdot k(x) = x\sin(1/x)$ for $x \neq 0$, then the effect of multiplying $k$ by $x$ is to decrease the magnitude of the oscillations as $x$ approaches zero.

The graph of $l$ is sandwiched between $|x|$ and $-|x|$, since $|x\sin(1/x)| \leq |x|$. Why? [**11.2**] If we define $l(0) = 0$, then do we obtain a function which is continuous at $x = 0$? The graph of $l$ in Figure 11.7 certainly suggests that it is.

(5) Let $m(x) = \sin x / x$ for $x \neq 0$.

  (a) Determine those $x$ for which $m(x) = 0$. [**11.3**]
  (b) Find curves between which the graph of $m$ oscillates. [**11.4**]
  (c) Sketch the graph of $m$ and think of a sensible value for $m(0)$. [**11.5**]

Our final two examples are much stranger than the ones we have just examined. The graphs are much harder to imagine and they appear to be a mess, from the point of view of intuitive ideas of continuity.

(6) Let $d(x)$ = the number of 7's in the decimal expansion of $x$ if this expansion is finite, and $d(x) = 0$ otherwise. First, $d(x) = 0$ for every irrational $x$. Why? [**11.6**] Also, $d(x) = n$ at infinitely many rationals in any interval of the real line for any $n \in \mathbb{N}$. For example, on $[0, 0.1]$, $d(x) = 4$ for $x$'s of the form $r/10, r/10^2, \ldots$, where $r = 0.7777$. Part of the graph of $d$ is rather clumsily shown in Figure 11.8. No point of this graph lies directly above any other, and yet it seems that the graph will ultimately consist of horizontal lines at levels 0, 1, 2, etc. However, there must be many gaps in every such line, so perhaps $d$ is discontinuous at every $x$!

(7) In a similar vein, consider

$$D(x) = \begin{cases} 0 & \text{if } x \text{ is irrational} \\ 1/q & \text{if } x = p/q \text{ is a rational in its lowest form} \end{cases}$$

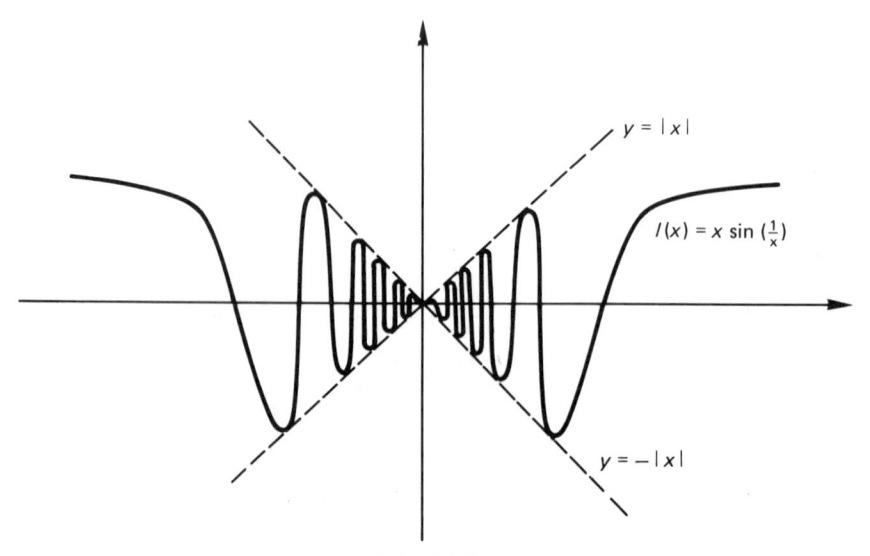

**Fig. 11.7**

## GRAPHS AND CONTINUITY

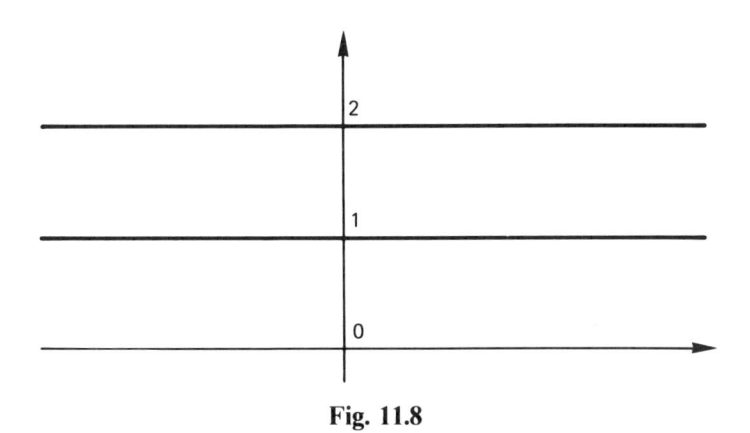

**Fig. 11.8**

The graph of D, called *Dirichlet's function*, consists of points variously splattered about the plane. To make life easier, we restrict the domain of D to be the real interval $(0, 1)$ and consider rationals with denominator 2, then 3, etc. Hence,

$D(x) = 0$     for all irrationals in $(0, 1)$
$D(x) = 1/2$     for $x = 1/2$
$D(x) = 1/3$     for $x = 1/3, 2/3$
$D(x) = 1/4$     for $x = 1/4, 3/4$ (NO, not $2/4$)

In fact, $D(x) = 1/q$ for at most $(q - 1)$ rationals in $(0, 1)$. For what values of $q$ does $D$ take the value $1/q$ precisely $(q - 1)$ times in $(0, 1)$? [**11.7**] The graph of $D$ appears as a pyramid of dots, as in Figure 11.9. As with the previous example, perhaps $D$ is everywhere discontinuous?

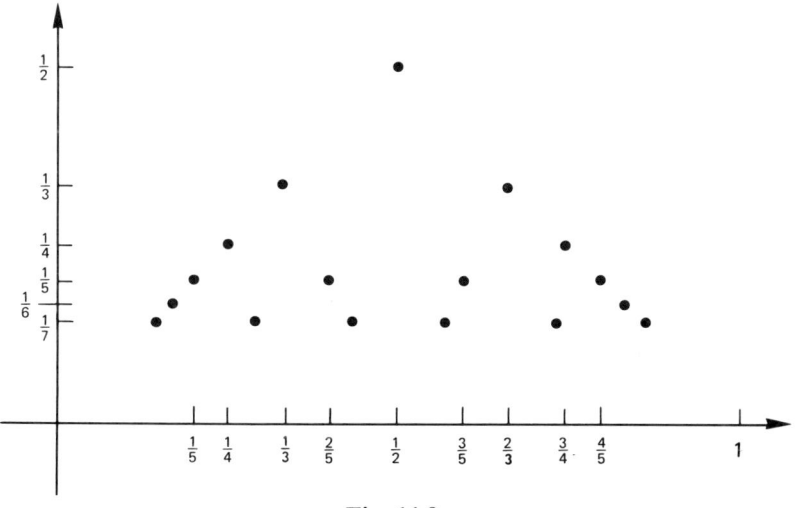

**Fig. 11.9**

167

In the next section we shall define continuity at a point. Our examples so far show why we need to concentrate on the behaviour of a function at (or, strictly speaking, arbitrarily close to) a fixed point. In example (2) we remarked that the graph of $y = x^2$ ought to be continuous, since it passed from a point $P_1 = (x_1, x_1^2)$ to a close-by point $P_2 = (x_2, x_2^2)$ when $x_1$ was close to $x_2$. Immediately then we wish to concentrate on some fixed point ($P_1$, say) and investigate the distance of points $P_2$ from $P_1$ when $|x_2 - x_1|$ is small. In example (3) we saw that obvious discontinuities occurred at isolated points and in example (4) the point $x = 0$ was of interest. In examples (6) and (7) all points seemed isolated (or are they?) Hence, we are drawn to define the notion of continuity at a point, with reference, of course, to all nearby points!

## 11.3 CONTINUITY

So to work. How do we capture the notion of a function $f$ being continuous at some fixed $x_0$ in its domain? Let's suppose that $f$ is defined for all $x$ in some open interval $I = (a, b)$ containing $x_0$. This means that $f(x_0)$ is defined and also $f(x)$ for all $x$'s within a certain (small) positive distance of $x_0$. In the introduction to this chapter we stated the intermediate value theorem—a result which we hope any function continuous on an interval will satisfy. One immediate consequence of that result, often called the *interval theorem*, which seems to capture the geometric idea of continuity is as follows.

> If a function $f$ is continuous on some closed bounded interval, then the image of that interval is itself a closed bounded interval.

See Figure 11.10. The proof of this theorem is a fairly easy deduction from

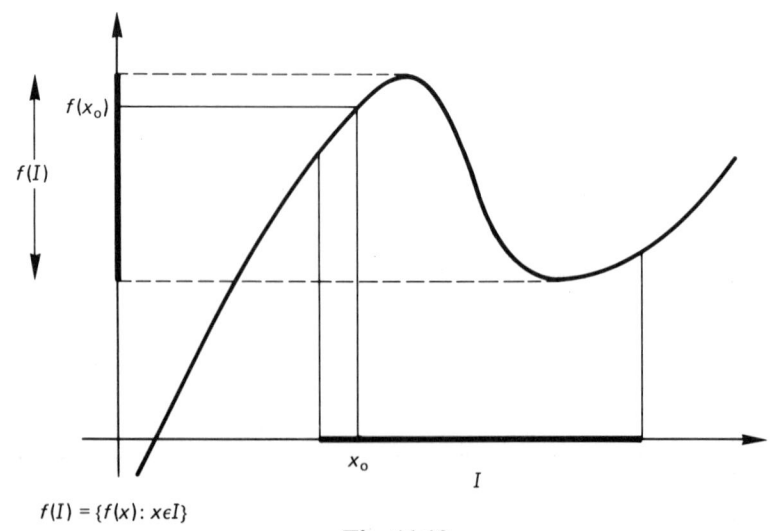

$f(I) = \{f(x): x \in I\}$

**Fig. 11.10**

the intermediate value theorem once we have shown that $f$ attains a greatest and least value on the interval. To achieve this, requires careful use of our forthcoming definition of continuity and our old friend the completeness axiom. There is so much to look forward to in an analysis course!

Intuitively, the smaller $I$ is, the smaller $f(I)$ should be in order that $f$ should map neighbouring $x$'s in the domain to neighbouring $f(x)$'s in the codomain. One very satisfactory definition of continuity exploits this idea by looking at the images of smaller and smaller intervals containing the fixed point $x_0$ and demanding that these images give a sequence of smaller and smaller intervals containing $f(x_0)$. Another, equivalent, definition considers sequences of real numbers with limit $x_0$ and demands that the image of such a sequence is a sequence in the codomain with limit $f(x_0)$. See Section 4.6 for a fuller discussion. Both these definitions and other variants are themselves equivalent to the more classical definition which we shall adopt. We concentrate on capturing the vague notion that when $x$ is close to $x_0$, $f(x)$ is close to $f(x_0)$. We turn the problem on its head and ask how close to $x_0$ $x$ must get before $f(x)$ is within a fixed (small) distance $\varepsilon$ of $f(x_0)$. The sort of answer we want for a continuous function is that there is a distance $\delta$ from $x_0$ such that *all* the $x$'s closer to $x_0$ than the discovered distance $\delta$ are all mapped to $f(x)$'s within the previously specified $\varepsilon$ distance of $f(x_0)$. This is illustrated in Figure 11.11.

So here goes:

---

**Official Definition of Continuity** A function $f$ is continuous at $x_0$ provided that for each $\varepsilon > 0$ (over which we have no control), $\exists \delta > 0$ ($\delta$ clearly depends on $\varepsilon$) such that $|f(x) - f(x_0)| < \varepsilon$ provided that $x$ satisfies $|x - x_0| < \delta$.

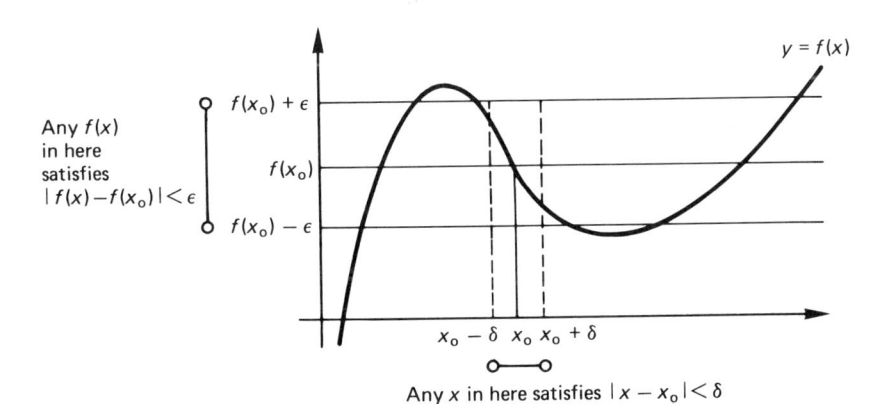

---

**Fig. 11.11**

So the game is this... someone specifies the $\varepsilon$; we have to find the $\delta$.

## Example (2) Revisited

Consider $g(x) = x^2$ and $x_0 = 1$. Now
$$|g(x) - g(x_0)| = |x^2 - 1|$$
Thus, $|g(x) - g(x_0)| < \varepsilon$ only when $|x^2 - 1| < \varepsilon$. But $x^2 - 1 = (x-1)(x+1)$ and so, if we can manufacture a suitable $\delta$ satisfying our definition, then, since $|x-1| < \delta$, $|x+1| = |x-1+2| < \delta + 2$, using the infamous triangle inequality. As a consequence, $|x^2 - 1| = |(x-1)(x+1)| < \delta(\delta + 2)$. So, all we need to arrange is for $\delta(\delta + 2) < \varepsilon$. To this end, we choose $\delta$ to be any positive real number not exceeding 1 or $\varepsilon/3$. Then, since $\delta \leqslant 1$, $\delta(\delta + 2) = \delta^2 + 2\delta \leqslant 3\delta$. Hence, $\delta(\delta + 2) \leqslant \varepsilon$, since $\delta \leqslant \varepsilon/3$. So, to summarise: with $g(x) = x^2$ and $x_0 = 1$, if we are given any $\varepsilon > 0$, we choose $\delta$ to be the minimum of 1 and $\varepsilon/3$; then $|x - x_0| < \delta \Rightarrow |g(x) - g(x_0)| < \varepsilon$. In other words, $g$ is continuous at $x_0 = 1$.

To obtain the continuity of $g$ at an arbitrary $x_0$ requires a similar argument and involves increased subtlety in order to determine the dependence of $\delta$ on $\varepsilon$.

## Shades of Example (3)

As a not too difficult exercise try proving that $f(x) = mx + c$, $m$ and $c$ constant, is continuous at any $x_0$. In other words, produce the $\delta > 0$ such that $|x - x_0| < \delta \Rightarrow |f(x) - f(x_0)| < \varepsilon$. [**11.8**] You have now shown that the function $h(x) = x - [x]$ is continuous for all non-integer values of $x_0$. But how does our definition fare at, say, $x_0 = 0$? Suppose that $h$ is continuous at $x_0 = 0$; then, for each $\varepsilon > 0$, $\exists \delta > 0$ such that $|x - 0| < \delta \Rightarrow |h(x) - h(0)| < \varepsilon$. Without loss of generality we can assume that $\delta < 1$; if some $\delta = \delta(\varepsilon)$ exists fulfilling our definition, then any smaller $\delta$ will also suffice. So we can write
$$|x| < \delta < 1 \Rightarrow |x - [x]| < \varepsilon$$
In particular, when $\varepsilon = 1/2$, $\exists \delta(1/2) = d < 1$ such that
$$|x| < d < 1 \Rightarrow |x - [x]| < 1/2 \qquad \dagger$$
But this implies that if $0 \leqslant x < d$, then $x < 1/2$ and so $(1 - x) > 1/2$. But $-d < -x \leqslant 0$ and so, using † again, we deduce that $(1 - x) < 1/2$. This contradiction means that our original assumption that $h$ was continuous at 0 was false. Hence, $h$ is not continuous at 0. Similar arguments can be used for any integer $x_0$.

## Example (4) Revisited

We do not use our definition of continuity to show that both $k(x) = \sin(1/x)$ and $l(x) = x \sin(1/x)$ are continuous for $x_0 \neq 0$. The details would be fiendishly difficult. In fact, except for the simplest of functions, our official definition is useless. What the definition is best used for is to establish certain general results which allow us to 'build up' more complicated continuous functions from simpler ones. The $\varepsilon$–$\delta$ proofs of these results are not as bad as they seem—it's all a matter of cooking up the $\delta$ to fit the $\varepsilon$. Typical of the sort of results we are talking about is that both the sum and product of two functions continuous at $x_0$ are themselves continuous at $x_0$. Using these facts, together with the continuity of $f(x) = mx + c$, prove that all polynomial functions are everywhere continuous. [**11.9**]

Our functions $k$ and $l$ require special attention at $x_0 = 0$.

(a) Consider $k(x) = \sin(1/x)$. We show rigorously that $k$ oscillates wildly as $x$ approaches 0. Hence, $k$ cannot settle down and be continuous at 0. For any $\delta > 0$, $\exists n \in \mathbb{N}$ such that $n > 1/2\pi\delta$. Why? [**11.10**] Hence, $(4n+1)\pi/2 > 4n\pi/2 > 1/\delta$, and so $0 < 2/(4n+1)\pi < 1/2n\pi < \delta$. So there exist $x$'s satisfying $|x| < \delta$, for any $\delta > 0$, where $k$ takes the values 0 and 1, respectively.

(b) Consider $l(x) = x \sin(1/x)$ and $l(0) = 0$. We examine the inequality $|l(x) - l(0)| < \varepsilon$. In other words, $|x \sin(1/x)| < \varepsilon$. Since $|\sin t| \leq 1$ for any $t \in \mathbb{R}$, we need only force $|x| < \varepsilon$. So, for any given $\varepsilon > 0$, choose $\delta = \varepsilon$. Then

$$|x - 0| < \delta \Rightarrow |x \sin(1/x) - 0| < \varepsilon$$

In other words, $l$ is continuous at 0. The $\delta$ used here is the same as the one you required for exercise [**11.8**]. By looking at the graph in Figure 11.7 perhaps you can see why—the graph of $l$ is squeezed between straight lines at $x_0 = 0$. Perhaps there is a general result to be proved here concerning the continuity at $x_0$ of a function squeezed, or sandwiched, between two functions known to be continuous at $x_0$? Indeed there is, but we shall resist the temptation here!

Let us now turn to the rather more pathological functions of examples (6) and (7). We concentrate on the latter and show first that Dirichlet's function

$$D(x) = \begin{cases} 0 & \text{is } x \text{ is irrational} \\ 1/q & \text{if } x = p/q \text{ is a rational in its lowest form} \end{cases}$$

is (perhaps not surprisingly) discontinuous at rational $x$'s in $(0, 1)$. So, let $x_0 = p/q \in (0, 1)$. Hence, $D(x_0) = 1/q$ and $D(x) > 1/(q+1)$ for at most $(q-1) + (q-2) + \cdots + 1 = q(q-1)/2$ points in $(0, 1)$. Explain why, using Figure 11.9 as a reference. [**11.11**] Now, for any $\delta > 0$ there are infinitely many $x$'s satisfying $|x - x_0| < \delta$. Hence, $D(x) \leq 1/(q+1)$ for infinitely many $x$'s in the interval $I = (x_0 - \delta, x_0 + \delta)$. But then

$$|D(x) - D(x_0)| \geq 1/q - 1/(q+1) = 1/q(q+1)$$

for all but a finite number of $x$'s in $I$. So, if we are given $\varepsilon < 1/q(q+1)$, we cannot manufacture a suitable $\delta$ satisfying the criteria for continuity. Hence, $D$ is discontinuous at rational $x$'s.

Now we consider an irrational $x_0$ in the interval $(0, 1)$. Can we find a $\delta > 0$ such that

$$|x - x_0| < \delta \Rightarrow |D(x) - D(x_0)| < \varepsilon?$$

Since $D(x_0) = 0$, we are trying to ensure that $|D(x)| < \varepsilon$. Choose $q \in \mathbb{N}$ such that $1/q < \varepsilon$. Why does such a $q$ exist? [**11.12**] There are only finitely many $x$'s (all rationals) such that $D(x) \geq 1/q$. Suppose that they are $x_1, x_2, \ldots, x_n$ and let $\delta$ be the minimum of $|x - x_1|, |x - x_2|, \ldots, |x - x_n|$. Then $\delta$ is the closest any of $x_1, x_2, \ldots, x_n$ get to $x_0$ and, since $x_0$ is irrational, it cannot equal any of $x_1, x_2, \ldots, x_n$. But now

$$|x - x_0| < \delta \Rightarrow |D(x) - D(x_0)| < 1/q < \varepsilon$$

We are forced to conclude that *D is continuous at all irrational x's*! What a bizarre example!

You could be forgiven for thinking that strange beasts like $D(x)$ are of no use. Once upon a time most mathematicians would have thought the same about continuous functions which are non-differentiable. For the purposes of the dying gasps of this book, interpret a differentiable function as one possessing a *smooth* continuous graph. By 'smooth' we mean that a tangent can be drawn at each point. It is difficult to imagine a continuous graph with a non-smooth outline (except perhaps at countably many points; see Figure 11.12). The function depicted takes the value 1 when $x = 1, 1/3, 1/5, \ldots$ and the value 0 when $x = 1/2, 1/4, 1/6, \ldots$. Its graph consists of straight-line segments joining successive points of the form $(1/2n, 0)$ and $(1/(2n-1), 1)$. The function is non-differentiable at all the corners. The mind boggles at what a continuous function which is nowhere differentiable looks like. Surely such exotic beasts are not the functions of real life! However, the newly

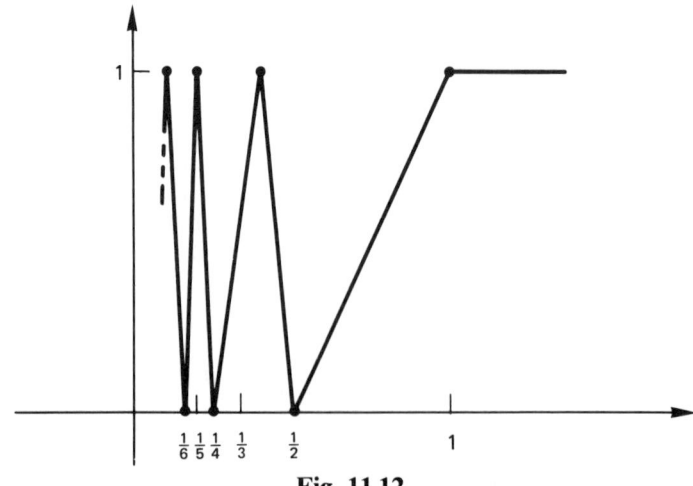

**Fig. 11.12**

emerging geometry of *fractals*, which we discuss in our concluding section, gives a lie to our previous remark.

## 11.4 FRACTALS

The French mathematician Benoit Mandelbrot published a paper in 1964 entitled; 'How long is the coastline of Britain?' (*Science*, volume 156, page 636). Mandelbrot was renovating an idea of the original-thinking Lewis Fry Richardson (1881–1953), who, in his mathematical investigations of the causes of war, became interested in a possible connection between the length of a border between two countries and the number of cross-border conflicts. It appeared to Richardson that the length of a border or a coastline increased without bound as the yardstick used to measure it was decreased in length. Indeed, the closer one observed a coastline the more wiggles appeared. Richardson essentially discovered the idea, crystallised by Mandelbrot, that curves such as the ones defining coastlines were in fact more than one-dimensional. They were 'between dimensions'. To be less vague, we construct a curve known as the *Koch snowflake*. Observe the sequence of shapes in Figure 11.13. To move from one shape to the next, we replace the 'initiator' •————————• by the 'generator'•—⁄\‾•. The limit curve is unbelievably wiggly. It will be continuous but will not possess a tangent at any point. Another example of generating a fractal curve is given in Figure 11.14. The generator in this case is

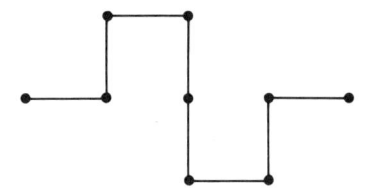

The limit curve is again the fractal curve required. At each stage of the process, notice that the area bounded is preserved in this example.

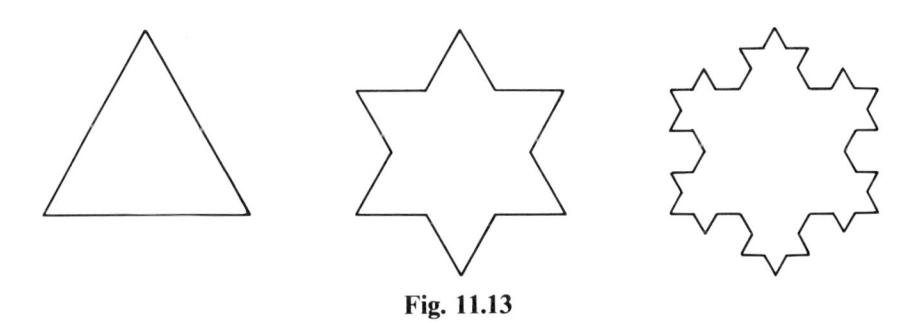

**Fig. 11.13**

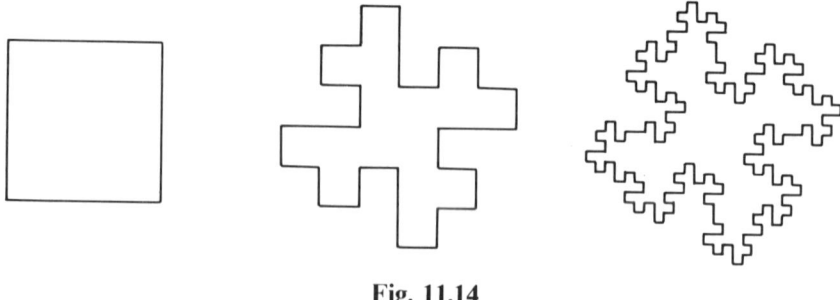

**Fig. 11.14**

The topological dimension of such curves is unity, since at each stage they are topologically equivalent to a circle. To obtain the *fractal dimension*, we let $N$ denote the number of parts of the basic generator and $R$ denote the similarity ratio. As illustrations: in Figure 11.13, $N = 4$ and $R = 3$; in Figure 11.14, $N = 8$ and $R = 4$. The fractal dimension, $D$, is given by

$$D = (\log_c N)/(\log_c R)$$

So, for the Koch snowflake, $D = (\log_c 4)/(\log_c 3) = 1.26$, which is roughly the fractal dimension of the coastline of Britain! The fractal generated by the process in Figure 11.14 has fractal dimension $D = 1.5$.

Fractals can be used to model many diverse areas of science and art. See, for example, Mandelbrot's definitive work, *The Fractal Geometry of Nature*. The current upsurge and advances in the sophistication of computer graphics allows us to investigate fractals more and more easily. Pictures such as a

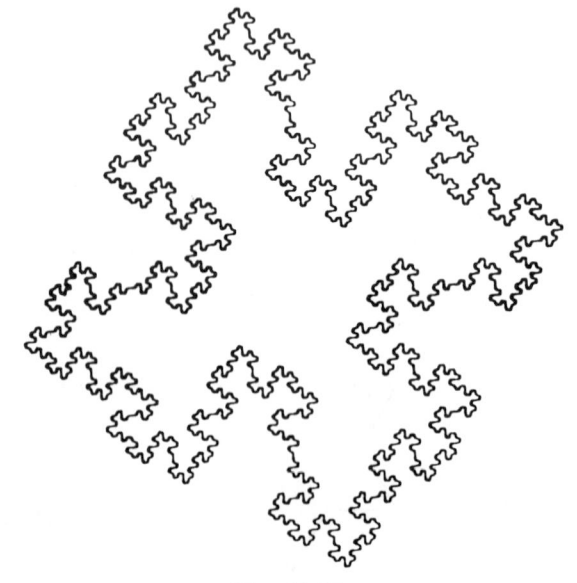

**Fig. 11.15**

later stage in the development of Figure 11.14, shown in Figure 11.15, are then indeed pale shadows of the real thing.

But is this fractal dimension at all meaningful? We conclude by looking at a fractal curve of dimension $D = 2$ (the topological dimension of the plane). Such curves were known to Peano and were in their day (1890) regarded as pathological or wild. They are eminently continuous curves which are everywhere non-differentiable and, more importantly, they are space filling. In Figure 11.16 the generator used is

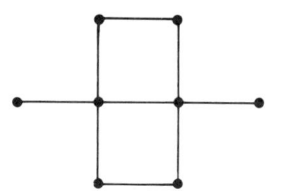

or, to show the path along the generator more clearly,

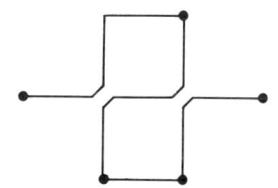

The process is repeated *ad infinitum*, to obtain the so-called *Cesaro curve*. Its fractal dimension is $D = (\log_e 9)/(\log_e 3) = 2$. The limit curve traverses every point of space! A later stage of the process is depicted in Figure 11.17.

We have indeed come a long way from our fumbling realisation that the real line contained more than just rationals. We have been forced into careful definitions of the real number line, infinity and continuity and many other concepts. We hope that some of the richness and excitement of analysis has

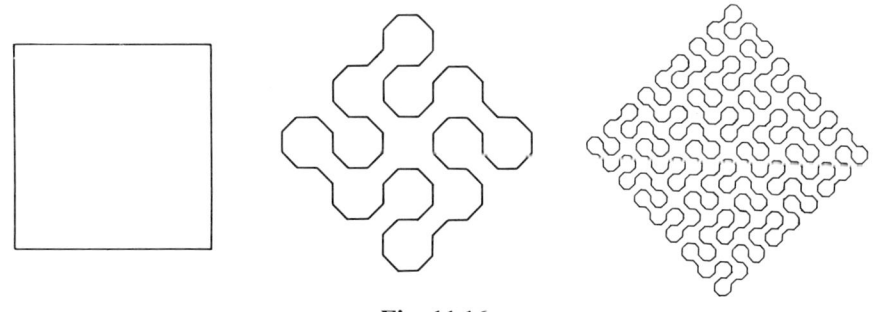

**Fig. 11.16**

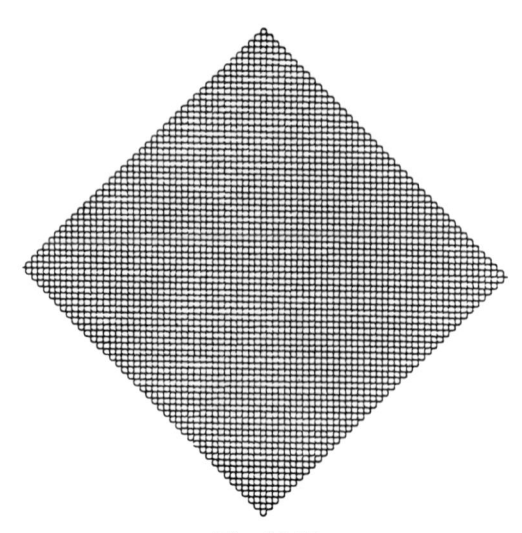

**Fig. 11.17**

been revealed to you. The rigour present in current mathematics is at worst a necessary evil but at best a thing of beauty to be caressed and respected. Mathematicians are notoriously bad at communicating the nature of their subject to the uninitiated—the authors feel that they are no exception to this rule. We hope that our book in some small way explains to you what 'grabs' us about mathematics, and that in your later career you, too, will gain the pleasure and enjoyment that mathematics has brought us. So long!

# Suggestions for Further Reading

## CHAPTER 1

E. T. Bell, *The Last Problem*, Gollancz (1962). The historical background to what is probably the most famous of all unsolved mathematical problems.

D. R. Hofstadter, *Gödel, Escher, Bach*: *An Eternal Golden Braid*, Penguin (1980). A real *pièce de résistance*, guiding us to a similar destination as Smullyan, but along a path strewn liberally with the music of Bach, the art of Escher and Lewis Carroll-type dialogues.

M. Jeger and M. Rueff, *Sets and Boolean Algebra*, Allen and Unwin (1970). Examines the relationship between the algebras of sets, switching circuits and probability, with some applications of these problems.

R. Smullyan, *The Lady or the Tiger?—and Other Logical Puzzles*, Pelican (1982). A highly entertaining sequence of logical puzzles leading gently to the heights of mathematical logic.

R. R. Stoll, *Sets, Logic and Axiomatic Theories*, 2nd edition, Freeman (1974). Analyses such concepts as validity of arguments and consistency of sets of statements, using the level of formal logic introduced in this chapter. It also connects with our Chapter 3 with a discussion of relations and with Chapter 8 in the properties of axiom systems.

## CHAPTER 2

D. M. Burton, *Elementary Number Theory*, Allyn and Bacon (1980). A more orthodox text than Burn; a very good introduction.

## CHAPTER 3

L. W. Cohen and G. Ehrlich, *The Structure of the Real Number System*, Van Nostrand (1962). Rather formal, but thorough. Fills the gaps in the path, $\mathbb{N} \to \mathbb{Z} \to \mathbb{Q} \to \mathbb{R}$, and also does not, as we have done, take $\mathbb{N}$ for granted, but derives its properties from Peano's axioms.

## CHAPTER 4

R. P. Burn, *A Pathway into Number Theory*, Cambridge University Press (1982). An excellent 'do-it-yourself' journey.

D. Wheeler, $\mathbb{R}$ *is for Real*, Open University Press (1974). Its main aim is to end the 'conspiracy of silence' concerning the jump from $\mathbb{Q}$ to $\mathbb{R}$, and has a similar philosophy to this book.

## CHAPTER 5

C. Cooke, *Exploring Graph Theory*, Parts 1 and 2, Keele Mathematical Education Publications (1985). These graphs are the vertex, edge variety and not the $y = x^2$ type! A very good learning-by-doing introduction.

G. Polya, *Mathematics and Plausible Reasoning*. Volume 1: *Induction and Analogy in Mathematics*, Princeton University Press (1954). Contains more insight into the topic of this chapter—and much more. A fascinating anthology round the theme of how mathematicians think (or should think!)

## CHAPTER 6

F. J. Budden, *The Fascination of Groups*, Cambridge University Press (1972). Long but easy to read. A very pleasant way to taste the flavour of abstract algebra—written by an enthusiastic, very professional amateur.

C. D. H. Cooper, *Permutations and Groups*, Murray (1975). Short, also easy to read, and puts permutations into the wider context of abstract algebra.

## CHAPTER 8

K. G. Binmore, *Logic, Sets and Numbers*, Cambridge University Press (1980). Relevant to our Chapter 1 and parts of Chapter 9. Concise and readable, with a good discussion of the axioms of $\mathbb{R}$.

## CHAPTER 9

J. W. Dauben, *Georg Cantor: His Mathematics and Philosophy of the Infinite*, Harvard University Press (1979). A very thorough account of Cantor's life and his mathematics.

N. Ya. Vilenkin, *Stories about Sets*, Academic Press (1968). A classic tale. An enjoyable way to learn the rudiments of Cantor's theory through this saga of the hotel with infinitely many rooms.

M. M. Zuckerman, *Sets and Transfinite Numbers*, Macmillan (1974). A good standard text to digest if inspired to go beyond our Chapter 9. Not for the faint-hearted.

## CHAPTER 10

K. G. Binmore, *Mathematical Analysis: A Straightforward Approach*, 2nd edition, Cambridge University Press (1982). One of the best beginner's standard texts for proper analysis.

A. Gardiner, *Infinite Processes: Background to Analysis*, Springer (1982). A leisurely but thorough analysis of analysis! If you think you understand limits, this book will show you that you don't—then provide you with a much more solid understanding.

J. A. Green, *Sequences and Series*, Routledge and Kegan Paul (1966). Packed full of examples of sequences and series.

H. E. Huntley, *The Divine Proportion*, Dover (1970). If your appetite for the golden ratio has been whetted by this chapter, Huntley's book goes a long way towards satisfying it.

## CHAPTER 11

C. W. Clark, *Mathematical Analysis*, Wadsworth (1982). An excellent introduction to real analysis. A carefully structured text with a wealth of interesting examples.

B. B. Mandelbrot, *The Fractal Geometry of Nature*, Freeman (1982). This unusual and beautifully illustrated book covers all the mathematics involved in fractal geometry. The author has a style of his own and the beauty and mystery of fractals is well presented together with historical and aesthetically pleasing anecdotes.

M. Spivak, *Calculus*, Benjamin (1967). Dedicated to Y. P. (yellow pig). A serious text which covers analysis very thoroughly but written with a light entertaining touch. Plenty of 'verbal garbage' to lift one's spirits!

# Hints and Solutions to Selected Exercises

[**1.1**] If 'I am lying' is a true statement, then I am telling a lie and so 'I am lying' is a false statement. Conversely, if 'I am lying' is a false statement, then I am telling the truth and so 'I am lying' is a true statement. Since mathematical statements are either true or false, we cannot allow 'I am lying' to be counted as a logical statement.

[**1.2**] It is not a Sunday, since the statements uttered by the two brothers cannot both be true. For other days of the week the brothers' statements have *opposite* truth values. Thus, if the first brother's statement is true, then he is Tweedledum and so the second brother is Tweedledee, which accords with the second brother's false statement. On the other hand, if the first brother is lying, then the second brother is Tweedledum, which accords this time with the second brother's true statement. Alice cannot tell which is which.

[1.3]

| P | Q | ~Q | P ∧ ~Q | ~(P ∧ ~Q) | P ⇒ Q |
|---|---|----|--------|-----------|-------|
| T | T | F  | F      | T         | T     |
| T | F | T  | T      | F         | F     |
| F | T | F  | F      | T         | T     |
| F | F | T  | F      | T         | T     |

[1.4]

| P | Q | ~P | ~Q | ~Q ⇒ ~P | P ⇒ Q |
|---|---|----|----|---------|-------|
| T | T | F  | F  | T       | T     |
| T | F | F  | T  | F       | F     |
| F | T | T  | F  | T       | T     |
| F | F | T  | T  | T       | T     |

[1.5] The conclusion contradicts the assertion that there exists a largest natural number and not the statement of the theorem, as claimed.

[1.6]  $A \cap (A \cup B)' = A \cap (A' \cap B')$   (De Morgan)

$\qquad\qquad\qquad = (A \cap A') \cap B'$   (Associativity)

$\qquad\qquad\qquad = \emptyset \cap B'$   (Complement)

$\qquad\qquad\qquad = \emptyset$   (Identity)

[1.7]  $(x, y) \in (A \times B) \cap (B \times A) \Leftrightarrow (x, y) \in (A \times B) \land (x, y) \in (B \times A)$

$\qquad\qquad\qquad \Leftrightarrow x \in A \land y \in B \land x \in B \land y \in A$

$\qquad\qquad\qquad \Leftrightarrow x \in A \cap B \land y \in A \cap B$

$\qquad\qquad\qquad \Leftrightarrow (x, y) \in (A \cap B) \times (A \cap B)$

$(x, y) \in (A \times B) \cup (B \times A) \Leftrightarrow (x, y) \in (A \times B) \lor (x, y) \in (B \times A)$

$\qquad\qquad\qquad \Leftrightarrow (x \in A \land y \in B) \lor (x \in B \land y \in A)$

$\qquad\qquad\qquad \Rightarrow (x \in A \lor x \in B) \land (y \in B \lor y \in A)$

$\qquad\qquad\qquad \Rightarrow x \in A \cup B \land y \in A \cup B$

$\qquad\qquad\qquad \Rightarrow (x, y) \in (A \cup B) \times (A \cup B)$

Not all of the implications can be reversed, as the following simple example shows: let $A = \{0\}$ and $B = \{1\}$; then $(A \cup B) \times (A \cup B)$ contains the four elements (0, 0), (0, 1), (1, 0) and (1, 1), whereas $(A \times B) \cup (B \times A)$ contains the two elements (0, 1) and (1, 0).

[1.8]

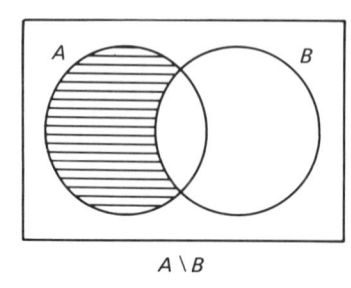

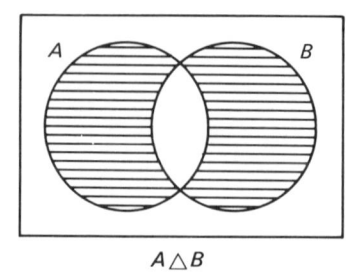

$A \setminus B$ ... $A \triangle B$

[2.1] If 10 780 and 39 102 had alternative prime factorisations, picking out the prime powers common to both could lead to something different from $2 \times 7^2$.

[2.2] Show that $(4n - 3) \times (4m - 3) = 4k - 3$ for some $k \in \mathbb{N}$.

[2.3] $1617 = 4 \times 405 - 3$.

[2.5] Find *primes* $a, b, c$, *not* in $S$. Show that $ab$, $c^2$, $ac$ and $bc$ *are* in $S$ and note that $ab \times c^2 = ac \times bc$.

[2.6] Read on.

[2.7] Otherwise $k$ would be prime and the *only* prime factorisation of $k$ would be $k$ itself.

[2.8] In the proof of UPF we needed to examine the natural number $k - pp'$ which was the difference between two natural numbers. If we tried to copy this proof in $S$, we would have to consider the difference between two $S$-numbers, which would not be an $S$-number.

[2.9] 7, because $7 = (1 + \sqrt{-6})(1 - \sqrt{-6})$.

[2.10] $10 = (2 + \sqrt{-6})(2 - \sqrt{-6}) = 2 \times 5$, and we show that $2, 5, 2 + \sqrt{-6}$ and $2 - \sqrt{-6}$ are all $T$-prime. Take 5 first: suppose that $5 = (a + b\sqrt{-6})(a - b\sqrt{-6})$ for some integers $a, b, c, d$. This implies $ac - 6bd = 5$, $bc + ad = 0$. Note that $a = b = 0$ is impossible, because this would mean $5 = 0$, so, solving for $d$, we get $d = -5b/(a^2 + 6b^2)d = 0 \Rightarrow b = 0 \Rightarrow 5 = ac \Rightarrow$ only the trivial solution by UPF in $\mathbb{N}$. So $d \neq 0$. Then

$$|d| = \frac{5|b|}{|a^2 + 6b^2|} \leqslant 5\left|\frac{b}{6b^2}\right| = \frac{5}{6|b|} < 1$$

But $d \neq 0$ and $|d| < 1$ is impossible, since $d$ is an integer. Hence, 5 is $T$-prime.

Now consider $2 + \sqrt{-6}$ and suppose this is $(a + b\sqrt{-6})(c + d\sqrt{-6})$. This leads to $ac - 6bd = 2$, $bc + ad = 1$, and, solving for $c$ and $d$, we obtain $c = (6b + 2a)/(a^2 + 6b^2)$ and $d = (a - 2b)/(a^2 + 6b^2)$. (Again $a^2 + 6b^2 \neq 0$, for otherwise $2 + \sqrt{-6} = 0$.)

If $a \neq 0$, $b = 0$, we get $d = 1/a \Rightarrow |a| = 1$, which are just the trivial solutions, $2 + \sqrt{-6} = \pm 1 \times \pm(2 + \sqrt{-6})$.

If $a = 0$, $b \neq 0$, we get $c = 1/b$, from which $|c| = |d| = 1$ and $d = -b/3$, which is impossible, since $d$ is integral.

182

If $d = 0$, then $a = 2b$, $c = 1/b$, so $c = b = \pm 1$, which again leads to only the trivial solutions.

Finally, if $a, b, d$ are all non-zero, we have

$$|d| = \left|\frac{a - 2b}{a^2 + 6b^2}\right| = \frac{|a - 2b|}{|a|^2 + 6|b|^2} \leq \frac{|a| + 2|b|}{|a|^2 + 6|b|^2} \leq \frac{|a| + 2|b|}{|a| + 6|b|} < 1$$

which is impossible, so $2 + \sqrt{-6}$ is $T$-prime.

Similarly, 2 and $2 - \sqrt{-6}$ can be shown to be $T$-prime.

[**2.11**] There is no way of defining $<$ on the complex numbers which is consistent with $+$ and $-$, but in the proof of UPF such consistency was required. We needed, for example, to be able to say that $k - pp'$ was less than $k$ because $pp'$ was greater than zero.

[**3.3**] Use the linearity of the stretch: let 1 on $S$ be vertically below $y$ on $L$ after the stretch. Then $x:1 = 1:y$, so $y = 1/x$.

[**3.4**] O (on $S$) stays fixed relative to $L$, so it can't be stretched to lie below 1 on $L$.

[**3.5**] Let $(a', b') \in [a, b]$ and $(c', d') \in [c, d]$, so that $a' + b = b' + a$ and $c' + d = d' + c$, and show that $a'c' + b'd' + bc + ad = b'c' + a'd' + ac + bd$.

[**3.6**] The zero integer is $[n, n]$ and the unit integer is $[n + 1, n]$.

[**3.7**] 
$$-z = [b, a], \quad -1 = [n, n + 1]$$

and

$$-1 \otimes z = [na + (n + 1)b, (n + 1)a + nb]$$
$$= [na + nb + b, na + nb + a] = [b, a] = -z$$

[**3.12**] The zero is $[0, 1]$; the unity is $[1, 1]$.

[**3.13**] $-r = [-a, b]$; $r^{-1} = [b, a]$ unless $r$ is zero.

[**3.14**] $ab > 0 \Rightarrow$ neither $a$ nor $b$ is zero, and they have the same sign. $(x, y) \in [a, b] \Rightarrow xb = ay \Rightarrow x$ and $y$ have the same sign, and neither is zero $\Rightarrow xy > 0$.

[**3.15**] Positivity.

[**3.17**]
$$(r \boxplus t) \trianglelefteq (s \boxplus t) \Leftrightarrow (s \boxplus t) \boxminus (r \boxplus t) \in \mathbb{Q}^+$$
$$\Leftrightarrow (s \boxplus t) \boxplus (-(r \boxplus t)) \in \mathbb{Q}^+$$
$$\Leftrightarrow (s \boxplus t) \boxplus ((-r) \boxplus (-t)) \in \mathbb{Q}^+$$
$$\Leftrightarrow s \boxplus t \boxplus (-r) \boxplus (-t) \in \mathbb{Q}^+ \Leftrightarrow s \boxplus (-r) \boxplus 0 \in \mathbb{Q}^+$$
$$\Leftrightarrow s \boxminus (-r) \in \mathbb{Q}^+ \Leftrightarrow s \boxminus r \in \mathbb{Q}^+ \Leftrightarrow r \trianglelefteq s$$

This argument uses some properties of $\mathbb{Q}$ which have not been 'officially' proved. Check that you are aware of the gaps, and fill them if you have the energy. They are all easy.

[**3.19**] Show that $(q \boxplus r) \boxtimes [1, 2]$ ($\frac{1}{2}(q + r)$ in 'ordinary' notation) is a suitable $s$.

[3.20] $[2a, 2b] = [a, b]$, but check that, in general, $[a+c, b+d] \neq [2a+c, 2b+d]$.

[3.22] Let P, Q and R be the points $(x_1, y_1)$, $(x_2, y_2)$ and $(x_3, y_3)$. Write down the conditions that the relationship holds between P and Q, Q and R, but *not* P and R, and show that they are incompatible.

[3.23] What if the cups have 20, 24 and 28 grains of sugar?

[3.24] 256. The truth table for a proposition with three statement letters has eight lines. The class of the proposition is determined by the assignments of T or F to each line, and there are $2^8 = 256$ such assignments—a choice of 2 (T or F) for each of the 8 lines.

[3.26] The 'proof' can only get started if there *is* a $b$ for which $a \mathcal{R} b$. If $\mathcal{R}$ is $>$ on $\mathbb{N}$ and $a$ is 1, there is no corresponding $b$.

[3.31] 5. Just count the ways of partitioning $\{a, b, c\}$.

[3.32] (i) $(b,b)$;  (ii) $(b,a), (c,a), (d,b)$;  (iii) $(a,d)$
       (i) $(b,b)$;  (ii) $(b,a), (c,a)$;         (iii) None.

[3.33] $\forall a, a^2 - a^2 = 0$, which is a multiple of $n$, so $a \alpha a$.

$$a^2 - b^2 = kn \Rightarrow b^2 - a^2 = -kn$$

so $a \alpha b \Rightarrow b \alpha a$.

$$a^2 - b^2 = kn \wedge b^2 - c^2 = ln \Rightarrow a^2 - c^2 = (k+l)n$$

so $a \alpha b \wedge b \alpha c \Rightarrow a \alpha c$. Hence, $\alpha$ is an equivalence relation.

For $n = 7$,

$$C_m = \{b \in \mathbb{N}: m^2 - b^2 = 7k \text{ for some integer } k\}$$

So $b \in C_1$ if and only if $1 - b^2$ is a multiple of 7. That is, $(1-b)(1+b)$ is a multiple of 7, so $b$ differs by 1 from a multiple of 7, so

$$C_1 = \{1, 8, 15, 22, 29, \ldots, 6, 13, 20, 27, \ldots\}$$

Similarly,

$$C_2 = \{2, 9, 16, 23, 30, \ldots, 5, 12, 19, 26, \ldots\}$$

$$C_3 = \{3, 10, 17, 24, 31, \ldots, 4, 11, 18, 25, \ldots\}$$

$$C_7 = \{7, 14, 21, 28, 35, \ldots\}$$

and these are the only equivalence classes.

For $n = 8$, $b \in C_1$ if and only if $1 - b^2$ is a multiple of 8. That is, $(1-b)(1+b)$ is a multiple of 8, which will be the case when $b$ is any odd number, for then one of $1 - b$ or $1 + b$ is a multiple of 4 and the other is even. Hence,

$$C_1 = \{1, 3, 5, 7, 9, 11, \ldots\}$$

Similarly,

$$C_2 = \{2, 6, 10, 14, 18, 22, \ldots\}$$

and
$$C_4 = \{4, 8, 12, 16, 20, 24, \ldots\}$$
and these are all the classes.

Finally, for $n = 10$ the distinct classes are
$$C_1 = \{10k + 9,\ 10k - 9: k \in \mathbb{N}\}$$
$$C_2 = \{10k + 8,\ 10k - 8: k \in \mathbb{N}\}$$
$$C_3 = \{10k + 7,\ 10k - 7: k \in \mathbb{N}\}$$
$$C_4 = \{10k + 6,\ 10k - 6: k \in \mathbb{N}\}$$
$$C_5 = \{5k: k \in \mathbb{N}\}$$

[**3.34**] $C_{64} = \{64\}$; $C_{32} = \{32, 96\}$; $C_{16} = \{16, 48, 80, 81\}$

$C_8 = \{8, 24, 40, 56, 72, 88, 27, 54\}$

$C_4 = \{4, 12, 20, 28, 36, 44, 52, 60, 68, 76, 84, 92, 100, 9, 18, 45, 63, 90, 99, 25, 50, 75, 49\}$

and the rest are in $C_2$.

[**3.35**] $\tau$ is reflexive and symmetric but transitivity fails. To see this, suppose $(a,b)\tau(c,d)$ and $(c,d)\tau(e,f)$. This means $c^2 b = a^2 d$ and $e^2 d = c^2 f$. If no coordinate is zero, we can deduce from these that
$$\frac{e^2 b}{a^2 f} = \frac{e^2 db}{a^2 fd} = \frac{c^2 fb}{a^2 fd} = \frac{c^2 b}{a^2 d} = 1$$
so $e^2 b = a^2 f$—i.e. $(a,b)\tau(e,f)$. However, if only $b$, $c$ and $d$ are zero, it is clear that $(a,b)\tau(c,d)$ and $(c,d)\tau(e,f)$, but $(a,b)$ is not related to $(e,f)$.

To see how $\tau$ partitions the non-axial points into equivalence classes,
$$C_{(a,b)} = \{(x,y): x^2 b = a^2 y,\ x \neq 0,\ y \neq 0\}$$
So $y = x^2 k$, where $k$ is the constant $b/a^2$—i.e. the equivalence classes are parabolas symmetric about the y-axis, through the origin.

[**3.37**] Suppose $A\mathcal{R}B$ and $B\mathcal{R}C$. Draw the usual Venn diagram of $A$, $B$ and $C$, and shade the regions now known to represent finite sets. To see that $A\mathcal{R}C$ note that $A\setminus C$ and $C\setminus A$ are unions of subsets of sets known to be finite, so are themselves finite. This establishes transitivity of $\mathcal{R}$. Symmetry and reflexivity are trivial.

If $S$ is taken to be $\mathbb{N}$, then $\mathbb{N}$ and $\mathbb{N}\setminus\{1\}$ are in the same class, but $\mathbb{N}$ and the set of even natural numbers are in different classes.

[**4.1**] Suppose that one collection contained $m$ unit masses and $n\sqrt{5}$ masses, and the other $p$ units and $q\sqrt{5}$ units. Then the ratio between their masses is $(m + n\sqrt{5})/(p + q\sqrt{5})$, so we want to show that this cannot be $\sqrt{7}$.

Suppose that it was: so

$$\sqrt{7} = \frac{m+n\sqrt{5}}{p+q\sqrt{5}} = \frac{(m+n\sqrt{5})(p-q\sqrt{5})}{(p+q\sqrt{5})(p-q\sqrt{5})}$$

which gives

$$\sqrt{7} = \frac{mp-5nq}{p^2-5q^2} + \sqrt{5}\cdot\frac{(np-mq)}{p^2-5q^2}$$

or $\sqrt{7} = r_1 + \sqrt{5}(r_2)$, where $r_1, r_2 \in \mathbb{Q}$. Squaring:

$$7 = r_1^2 + 5r_2^2 + 2r_1 r_2 \sqrt{5}$$

which gives

$$\sqrt{5} = \frac{7 - r_1^2 - 5r_2^2}{2r_1 r_2}$$

—a rational number!

[4.2] They are all equiangular by construction and the sides which are parallel to PQ are equal.

[4.4] Each member of $M$ is of the form $(2^n - 1)/2n$ for $n \in \mathbb{N}$. Whichever member you propose as the biggest, say $(2^k - 1)/2k$, the next one, $(2^{k+1} - 1)/2^{k+1}$, is bigger.

[4.5] If $x$ is a common measure of AB and CD, then so are $x/2, x/3, x/4, \ldots$.

[4.6] Any length $x$ which 'measures' AB is of the form $AB/n$. If any longer measure exists, it is of the form $AB/m$ for $m < n$. Hence, the set of 'measures' of AB at least as long as $x$ is a subset of the *finite* set,

$$\left\{\frac{AB}{n}, \frac{AB}{n-1}, \frac{AB}{n-2}, \ldots, \frac{AB}{1}\right\}$$

which necessarily has a greatest member.

[4.8] Show that any common factor of $a$ and $b$ is a common factor of $b$ and $a - kb$, and vice versa. Hence, the *set* of common factors of $a$ and $b$ is equal to the set of common factors of $b$ and $a - kb$. Hence, their greatest elements must be equal.

[4.9] Consider any two consecutive steps of the process. The first divides $a$ by $b$, to give a remainder of $c$, so $b > c$. The next divides $b$ by $c$ to give a remainder of $d$, so $c > d$. So the remainders at successive steps form a strictly decreasing sequence of natural numbers which necessarily terminates in a finite number of terms.

[4.11] Just a matter of checking that $I_{k+1} \cap J_{k+1} \subseteq I_k \cap J_k$. This follows, because $x \in I_{k+1} \cap J_{k+1} \Rightarrow x \in I_{k+1}$ and $x \in J_{k+1} \Rightarrow x \in I_k$ and $x \in J_k \Rightarrow x \in I_k \cap J_k$.

[4.12] If the two nests converged to different points, say $p$ and $q$, we could place disjoint intervals $P$ and $Q$ round these points, and there would be some $k$ in $\mathbb{N}$ for which $I_k \subset P$ and $J_k \subset Q$. That is, $I_k \cap J_k = \emptyset$.

[4.13] If $p^2/q^2 < 5$, where $p, q \in \mathbb{N}$, show that $\exists n \in \mathbb{N}$ such that $((np+1)/nq)^2 < 5$.

[4.14] If $x$ is in the set, so is the larger number, $x+1$, and the smaller number, $\frac{1}{2}(1+x)$.

[4.15] Use the approach of [4.13].

[5.1] At the start of month $n$ suppose that the numbers of young, middle-aged and old pairs are $y_n$, $m_n$ and $o_n$, respectively (so $f_n = y_n + m_n + o_n$). At the start of the next month the young will have become middle-aged, the middle-aged old, and the $m_n + o_n$ old pairs now present will have produced $m_n + o_n$ young pairs. So, given the distribution at the start of any month, the distribution (and, hence, total) in any subsequent month may be calculated.

[5.2] Doing the calculation suggested above for three successive months:

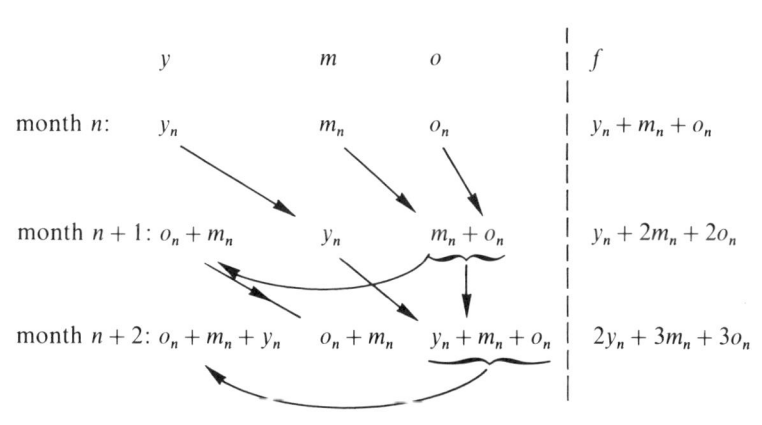

|  | y | m | o | f |
|---|---|---|---|---|
| month $n$: | $y_n$ | $m_n$ | $o_n$ | $y_n + m_n + o_n$ |
| month $n+1$: $o_n + m_n$ | | $y_n$ | $m_n + o_n$ | $y_n + 2m_n + 2o_n$ |
| month $n+2$: $o_n + m_n + y_n$ | | $o_n + m_n$ | $y_n + m_n + o_n$ | $2y_n + 3m_n + 3o_n$ |

—from which it is easy to see that $f_{n+2} = f_{n+1} + f_n$.

[5.6] $f(n) = 4^n - 3n - 1$; $f(1) = 0$, $f(2) = 9$, $f(3) = 54$, $f(4) = 243$, $f(5) = 1008$, $f(6) = 4077$. Conjecture. $S_n$: $f(n)$ is a multiple of 9 for all $n \in \mathbb{N}$. $S_1$ is clearly true. Assume $S_k$—i.e. $4^k - 3k - 1$ is a multiple of 9. Then $4^{k+1} - 3(k+1) - 1 = 4(4^k - 3k - 1) + 9k$, which, using the induction hypothesis, is divisible by nine.

[5.9] You should find that $2^n - 1$ is divisible by 3 when $n$ is even and not otherwise. $S_n$: $2^{2n} - 1$ is divisible by 3 for all $n \in \mathbb{N}$. $S_1$ is clearly true, and if $S_k$ holds, then $2^{2(k+1)} - 1 = 4(2^k - 1) + 3$, which is also divisible by 3. [Or try a direct proof by factorising $2^{2n} - 1$.]

[5.10] $S_1$ is true, and if $l = 1$, then $S_{l-1}$ wouldn't exist.

[5.11] $0 < 3 - \sqrt{5} < 1$, so $0 < k < l$. Also $(3 - \sqrt{5})l = 3l - n$, which is an integer.

[5.12] $k\sqrt{5} = (3 - \sqrt{5})l\sqrt{5} = (3nl/l) - 5l = 3n - 5l$.

[5.13] Work on $(10 - \sqrt{94})l$.

[5.14] There is no $n \in \mathbb{N}$ such that $0 < n - \sqrt{49} < 1$.

[5.20] Assuming $(1+x)^k \geqslant kx$,
$$(1+x)^{k+1} = (1+x)(1+x)^k \geqslant (1+x)kx = (k+kx)x$$
but what is required is $(1+x)^{k+1} \geqslant (k+1)x$, so our result is not good enough, since $kx$ is not necessarily $\geqslant 1$. Try proving the stronger result, $(1+x)^k \geqslant 1 + kx$, instead.

[5.21] The following argument derives a contradiction from denying $(\forall n)(S_n)$: $\sim (\forall n)(S_n) \Rightarrow (\exists n)(\sim S_n)$, so suppose $S_m$ is false. By $I_1''''$ $\exists l > m$ for which $S_l$ is true. Then, by repeated application of $I_2''''$, $S_m$ is true!

[5.22] This time we need to show that $k+1$ is fs if $k$ is fs. To do this, show that $k$'s representation can always be arranged to omit 1 or contain 1 but not 2.

[5.27] Suppose that $e + c \geqslant v$ is true for all graphs with $c$ at most $c_0$. This is true for $c_0 = 1$ by [5.26]. Let $G$ be a graph with $e$ edges, $v$ vertices and $c_0 + 1$ components. Think of $G$ as two separate graphs: $G_1$ with $e_1$, $v_1$, 1 edges, vertices and components, respectively, and $G_2$ with corresponding parameters $e_2$, $v_2$ and $c_0$. Then, by our supposition (induction hypothesis), for $G_1$:
$$e_1 + 1 \geqslant v_1$$
and for $G_2$:
$$e_2 + c_0 \geqslant v_2$$
Adding,
$$(e_1 + e_2) + (c_0 + 1) \geqslant (v_1 + v_2)$$
which is the result required for $G$.

[5.28] If a graph has only one vertex, it necessarily has no edges and one component, so the result clearly holds. Any graph of $v+1$ vertices may be thought of as a graph of $v$ vertices to which an extra vertex, possibly with incident edges, has been added. Now just consider what can happen to $e$, $v$ and $c$ as a result of adding the extra vertex and edges.

[5.29] Follow the method of the text but let $k+1$ be the number of letter-occurrences in $\mathscr{F}$. in $F$ and $G$ each have a score $\leqslant k$.

[5.30] If $k+1$ is the number of distinct letters in $\mathscr{F}$, then $F$ and $G$ could also each have $k+1$ distinct letters, so the induction hypothesis cannot be applied to them.

[5.31] Using the hint and the previous discussion of the general associative law (in Section 5.3.2), any such formula can be written as $\mathscr{F}_1 \Leftrightarrow \mathscr{F}_2 \Leftrightarrow \mathscr{F}_3 \Leftrightarrow \ldots \Leftrightarrow \mathscr{F}_n$ where $\mathscr{F}_i$ is $P_i \Leftrightarrow P_i \Leftrightarrow \ldots \Leftrightarrow P_i$. It is easy to check that if each $\mathscr{F}_i$ contains an even number of $P_i$'s, then it is a tautology, and that any number of tautological formulae joined by $\Leftrightarrow$ makes another tautology. Conversely, suppose that at least one of the $\mathscr{F}$'s, say $\mathscr{F}_i$, contains an odd number of $P_i$'s. Check that if $P_i$ is assigned the truth value F, then $\mathscr{F}_i$ is F. Finally, assign truth value T to all the other $P_j$ and check that this makes all the $\mathscr{F}_j$ (other than $\mathscr{F}_i$) true, and the whole formula then becomes false, so is not a tautology.

[5.32] The value is in fact prime for $x = 0, 1, 2, 3, \ldots 39$. You should see easily that it must be composite for $x = 41$, and, by thinking harder, that it is also composite for $x = 40$.

[5.33] For $n = 1, 2, 3, 4, 5$, the maximum number $r$ of regions is 1, 2, 4, 8, 16, but the pattern is *not* $r = 2^{n-1}$. It fails for $n = 6$, where $r = 31$.

[5.34] Any that you are likely to try will only have irreducible factors with coefficients $+1$ or $-1$. Amazingly, you have to go as far as $x^{105} - 1$ to find a counter-example—a gem undiscovered until 1941, by V. Ivanov. Here is the rogue factor:

$$x^{48} + x^{47} + x^{46} - x^{43} - x^{42} - 2x^{41} - x^{40} - x^{39} + x^{36} + x^{35} + x^{34} + x^{33}$$
$$+ x^{32} + x^{31} - x^{28} - x^{26} - x^{22} - x^{20} + x^{17} + x^{16} + x^{15} + x^{14} + x^{13}$$
$$+ x^{12} - x^9 - x^8 - 2x^7 - x^6 - x^5 + x^2 + x + 1$$

[5.35] The algebra is perfectly correct! *If* $1 + 2 + 3 + \cdots + k$ is equal to $\frac{1}{2}(k + 1/2)^2$, it does follow that $1 + 2 + 3 + \cdots + k + (k + 1)$ is equal to $\frac{1}{2}((k + 1) + 1/2)^2$, but this *does* not complete the proof because the 'if' clause above never is true. This is why checking $S_1$ is vital in order to get the induction started.

[5.36] This time $S_1$ has been checked. There is just a tiny hole in the argument: the deduction of $S_{k+1}$ from $S_k$ works for every $k$ *except 1* (a set of two elements can't be split into subsets $A$ and $B$ with an element in each of $A \backslash B$, $B \backslash A$ and $A \cap B$!)

[5.37] The first only is Hamiltonian.

[5.38] Otherwise the Hamiltonian circuit could be joined to $v$ by at most $n/2$ edges.

[5.39] The proof can only be valid if every graph of $n + 1$ vertices which satisfies the edge number condition *can* be constructed by adjoining an extra vertex and edges to an $n$-vertex graph which satisfies the edge number condition. A very simple counter-example to show that this is not the case is the single circuit of four vertices. This clearly satisfies the condition and has a Hamiltonian circuit, but it cannot be constructed from a 3-vertex graph satisfying the condition by the method required in the 'proof'.

[6.1] Show that (i) if $b \in A$, $\exists a \in A$ such that $(f \circ g)(a) = b$; (ii) if $(f \circ g)(a) = (f \circ g)(b)$, then $a = b$.

[6.3] First make a transposition which puts the correct element in the left-hand end position. Then put the correct element in the next place, $\ldots$, and so on.

189

**[6.4]** This example shows that a transposition can be expressed as a 'product' of two permutations neither of which is the identity. Hence, if the identity permutation is the analogue of the number 1, transpositions are not good analogues for primes.

**[6.5]** These two permutations are different, showing that, unlike multiplication of numbers, composition of permutations is not commutative.

**[6.6]**

$$\bullet \cdots \bullet \underbrace{\overset{a}{\underset{\downarrow}{\bullet}} \cdots \overset{b}{\underset{\downarrow}{\bullet}}}_{n \text{ elements}} \cdots \bullet$$

If elements $a$ and $b$ are transposed, and there are $n$ elements between them, then the pairs whose order is reversed are those consisting of $a$ and each of the $n$ elements between $a$ and $b$, those consisting of $b$ and each of the $n$ elements between $a$ and $b$, and the pair $ab$ itself—a total of $n+n+1 = 2n+1$ pairs.

**[6.7]** This is easy, really! The difficulty is in putting the explanation into words. Suppose that we start with a number of symbols in their 'natural' order, abcde...z, and make a series of transpositions. At each stage we look at each pair of symbols and note whether it is in its natural (n) order or reversed (r) order. Suppose the $n$th transposition causes $k_n$ pairs to switch from (n) to (r) and $l_n$ pairs to switch from (r) to (n). Note from **[6.6]** that $k_n + l_n$ is *odd* for each $n$. Let $t_n$ be the total number of r pairs after $n$ transpositions. Then $t_n = t_{n-1} + k_n - l_n$, and since $k_n + l_n$ is odd, the same is true of $k_n - l_n$, and so we have shown that $t_n$ and $t_{n-1}$ differ by an *odd* number. Finally, since $t_1$ is odd (by **[6.6]**), the parity of $t_n$ is the same as that of $n$, as required.

**[6.8]** Let $a_1, a_2, \ldots, a_n$ be the $n$ letters. Decide the position of $a_1$. This can be done in $n$ ways. Then fix $a_2$: $n-1$ choices for this as any one of the remaining $n-1$ places may be chosen. Hence, we have $n \times (n-1)$ ways of deciding the positions of $a_1$ and $a_2$. Continue with this line of argument.

**[6.9]** With $n=1$, there is necessarily only one permutation and this is necessarily the identity—an even permutation.

**[6.10]** Obvious because $p$ and $q$ shift disjoint sets of symbols, so $p$ and $q$ act independently.

**[6.11]** Suppose that $t_i$ is the first member of $(t_1, t_2, \ldots, t_{l-1})$ which is in $(s_1, s_2, \ldots, s_{k-1})$. Now $i > 1$ because of the way $t_1$ was chosen. Suppose $t_i = s_j$. Then we have $t_{i-1} \overset{p}{\to} t_i$ and $s_{j-1} \overset{p}{\to} t_i$ and $s_{j-1} \neq t_{i-1}$, so $p$ would not be one-to-one.

**[6.12]** (A G E K)∘(B J I)∘(C)∘(D M L F N)∘(H)

**[6.13]** An example is **[6.4]**: (A L) = (A C I)∘(A L I C).

**[6.14]** $(p_1 \circ p_2 \circ p_3) \circ (p_3^{-1} \circ p_2^{-1} \circ p_1^{-1}) = (p_1 \circ p_2) \circ (p_3 \circ p_3^{-1}) \circ (p_2^{-1} \circ p_1^{-1})$
$= (p_1 \circ p_2) \circ (p_2^{-1} \circ p_1^{-1}) = p_1 \circ (p_2 \circ p_2^{-1}) \circ p_1^{-1} = p_1 \circ p_1^{-1} = i$. So $p_3^{-1} \circ p_2^{-1} \circ p_1^{-1}$ must be the inverse of $p_1 \circ p_2 \circ p_3$.

**[6.15]** Zero.

**[6.16]** 1 and $-1$.

# HINTS AND SOLUTIONS TO SELECTED EXERCISES

A permutation on $n$ ($>1$) symbols is self-inverse if and only if, when expressed in disjoint cycle form, the cycles are all of length 2 (and 1).

[**6.17**] $p^{-1} = (a_m, a_{m-1}, a_{m-2}, \ldots, a_1)$.

[**6.18**] (i) and (ii) but not (iii)—except for special $p$ and $q$.

[**6.19**] There are only a finite number of permutations on $n$ symbols, so $p, p^2, p^3, \ldots$ cannot all be different. So suppose $p^n = p_m$ ($n > m$). Hence, $p^{n-m} = i$.

[**6.20**] The length is the same as the order.

[**6.21**] The order of $p$ is the least common multiple of the lengths of the $c_i$'s.

[**6.22**] If the cycle has even length, its parity is odd, and if the length is odd, the parity is even, because, for example, (a b c d e f g) = (a g)∘(a f)∘(a e)∘(a d)∘(a c)∘(a b).

[**6.24**] $p \sim p$ because $p = p^i$.

$$p \sim q \Rightarrow p = r \circ q \circ r^{-1} \Rightarrow q = r^{-1} \circ p \circ r = p^{r^{-1}} \Rightarrow q \sim p$$

$$p \sim q \wedge q \sim s \Rightarrow p = r \circ q \circ r^{-1} \wedge q = t \circ s \circ t^{-1}$$

$$\Rightarrow p = r \circ t \circ s \circ t^{-1} \circ r^{-1} = (r \circ t) \circ s \circ (r \circ t)^{-1}$$

$$= s^{r \circ t} \Rightarrow p \sim s$$

[**6.25**] $q^r(r(A)) = (q^r \circ r)(A) = (r \circ q \circ r^{-1} \circ r)(A)$

$$= (r \circ q \circ i)(A) = (r \circ q)(A) = r(q(A)) = r(B)$$

[**6.26**] The distinct cycle structures correspond to the equivalence classes. These are listed below for a set of five elements, together with the number of permutations in each class.

| | |
|---|---:|
| (.....) | 24 |
| (....)(.) | 30 |
| (...)(.)(.) | 20 |
| (..)(.)(.)(.) | 10 |
| (.)(.)(.)(.)(.) | 1 |
| (..)(...) | 20 |
| (..)(..)(.) | 15 |
| | 120 |

[**7.1**] Let $I = [a, b]_Q$ and $J = [c, d]_Q$. Now,

$$x + y \in I + J \Rightarrow x \in I \text{ and } y \in J$$

$$\Rightarrow a \leqslant x \leqslant b \text{ and } c \leqslant y \leqslant d$$

$$\Rightarrow a + c \leqslant x + y \leqslant b + d$$

$$\Rightarrow x + y \in [a + c, b + d]_Q$$

Conversely,

$$t \in [a+c, b+d]_\mathbb{Q} \Rightarrow a+c \leqslant t \leqslant b+d$$
$$\Rightarrow 0 \leqslant t-(a+c) \leqslant (b+d)-(a+c)$$
$$\Rightarrow 0 \leqslant t-(a+c) \leqslant (b-a)+(d-c)$$
$$\Rightarrow t-(a+c) = r_1 + r_2, \text{ where}$$
$$0 \leqslant r_1 \leqslant b-a \text{ and } 0 \leqslant r_2 \leqslant d-c$$
$$\Rightarrow t = (r_1+a)+(r_2+c), \text{ where}$$
$$a \leqslant r_1+a \leqslant b \text{ and } c \leqslant r_2+c \leqslant d$$
$$\Rightarrow t \in I+J$$

Similar arguments work for $I \times J$ and $-I$.

**[7.2]** Let $I = [a,b]_\mathbb{Q}$, where $a$ and $b$ have the same sign, so that 0 is excluded from $I$. For the sake of argument, we take $a$ and $b$ to be positive.

$$t \in 1/I \Leftrightarrow t = 1/x, \text{ where } a \leqslant x \leqslant b$$
$$\Leftrightarrow 1/b \leqslant t \leqslant 1/a$$
$$\Leftrightarrow t \in [1/b, 1/a]_\mathbb{Q}$$

**[7.3]** See solution **[7.1]** above.

**[7.4]** $[p,q]_\mathbb{Q} \times [p',q']_\mathbb{Q} = [pp', qq']_\mathbb{Q}$

**[7.5]** We show that if $\{I_n\}$ and $\{J_n\}$ are rational nests where all the intervals involved have positive rational end-points, then $\{I_n \times J_n\}$ is also a rational nest. More ingenuity is required for arbitrary rational nests.

If $I_n$ and $J_n$ are closed rational intervals, then so too is $I_n \times J_n$. Now,

$$xy \in I_{n+1} \times J_{n+1} \Rightarrow x \in I_{n+1} \text{ and } y \in J_{n+1}$$
$$\Rightarrow x \in I_n \text{ and } y \in J_n, \text{ since } \{I_n\} \text{ and } \{J_n\} \text{ are nests}$$
$$\Rightarrow xy \in I_n \times J_n$$

So criterion (i) for a rational nest holds. Now let $I_n = [p_n, q_n]_\mathbb{Q}$ and $J_n = [p_n', q_n']_\mathbb{Q}$, where $p_n, q_n, p_n'$ and $q_n'$ are all positive rationals. For each $m \in \mathbb{N}$, $\exists r_m \in \mathbb{Q}$ such that $2^{r_m} \geqslant q_m' + p_m$. Now, for any $n \in \mathbb{N}$, $\exists m \in \mathbb{N}$ such that $|I_m| \leqslant 1/2^{n+r_m}$ and $|J_m| \leqslant 1/2^{n+r_m}$. Finally,

$$|I_m \times J_m| = q_m q_m' - p_m p_m', \text{ using the formula in [7.4]}$$
$$= (q_m - p_m)q_m' + (q_m' - p_m')p_m$$
$$= |I_m|q_m' + |J_m|p_m$$
$$\leqslant (q_m' + p_m)/2^{n+r_m} \leqslant 1/2^n$$

thus verifying criterion (ii) for a rational nest.

**[7.6]** $\qquad \{I_n\} \times \{1/I_n\} = \{I_n \times 1/I_n\}$

and without loss of generality no $I_n$ contains 0. Also,
$$I_n \times 1/I_n = \{x \cdot 1/y : x, y \in I_n\}$$
Now, for each $x \in I_n$, we have $1/x \in 1/I_n$ and so $x \cdot 1/x \in I_n \times 1/I_n$. Hence, 1 is contained in every interval of the nest $\{I_n\} \times \{1/I_n\}$ and so this nest defines the unique real number 1.

[7.7] Follows immediately from the well-definedness of multiplication and the following definition of division of rational nests:
$$\{I_n\}/\{J_n\} = \{I_n\} \times \{1/J_n\}$$
where, as usual, no $J_n$ contains 0.

[7.8] We verify A2 and A8.

A2: $\{I_n\} + \{J_n\} = \{I_n + J_n\}$
$\phantom{A2: \{I_n\} + \{J_n\}} = \{J_n + I_n\}$, using commutativity of rationals
$\phantom{A2: \{I_n\} + \{J_n\}} = \{J_n\} + \{I_n\}$

A8: If $\{I_n\}$ defines a non-zero real number, then $\{1/I_n\}$ exists. By [7.6] $\{I_n\} \times \{1/I_n\}$ defines the real number 1.

[7.9] We prove A13 as an illustration. Since $a = b$ implies $a + c = b + c$ for any real numbers $a$, $b$ and $c$, it suffices to prove that
$$a < b \Rightarrow a + c < b + c$$
Let $\{I_n\}$, $\{J_n\}$ and $\{K_n\}$ be rational nests defining $a$, $b$ and $c$, respectively. Since $a \neq b$, then, for some $m \in \mathbb{N}$, the rational closed intervals $I_m$ and $J_m$ are such that $r < s$ for all $r \in I_m$ and for all $s \in J_m$. For any $t \in K_m$, we have by the order relation on $\mathbb{Q}$ that $r + t < s + t$. But this means that $I_m + K_m < J_m + K_m$ and, hence, $a + c < b + c$, as required.

[8.1] Now $0 + 0 = 0$ by A3. But $0 + (-0) = 0$ by A4 and, moreover, $(-0)$ is unique. Hence, $0 = -0$.

[8.2] For non-zero $x$, $(-x) \cdot (-x)^{-1} = 1$ by A8. Hence,
$$-(x^{-1}) \cdot ((-x) \cdot (-x)^{-1}) = -(x^{-1}) \text{ by A7}$$
So
$$(-(x^{-1}) \cdot (-x)) \cdot (-x)^{-1} = -(x^{-1}) \text{ by A5}$$
So, by Theorem 8.3,
$$(x^{-1} \cdot x) \cdot (-x)^{-1} = -(x^{-1})$$
Now, using A6, A7 and A8, $(-x)^{-1} = -(x^{-1})$. This proves Theorem 8.4.

For non-zero $x$, $x^{-1}$ exists. Moreover, $x^{-1} \neq 0$, since, if $x^{-1} = 0$, then, by Theorem 8.1, $x \cdot x^{-1}$ would equal 0. But A8 says that $x \cdot x^{-1} = 1$ and, hence, $0 = 1$, which contradicts A7. Since $x^{-1}$ is non-zero, $(x^{-1})^{-1}$ exists. Now, freely using the axioms of arithmetic,
$$x = x \cdot 1 = x \cdot (x^{-1} \cdot (x^{-1})^{-1}) = (x \cdot x^{-1}) \cdot (x^{-1})^{-1} = 1 \cdot (x^{-1})^{-1} = (x^{-1})^{-1}$$

This proves Theorem 8.5.

[**8.3**] If $-1 = 0$, then $-(-1) = -0$, since additive inverses are unique (A4). But $-(-1) = 1$, as we saw in the proof of Theorem 8.3 and $-0 = 0$, by exercise [**8.1**]. Hence, $1 = 0$, which contradicts A7.

[**8.4**] By Theorem 8.6, $0 < 1$, and $(-1)$ exists by A4. Hence, $0 + (-1) \leqslant 1 + (-1)$ by A13. Hence, $(-1) \leqslant 0$, using A2, A3 and A4. If $-1 = 1$, then $1 \leqslant 0$, which contradicts $0 < 1$. Hence, $-1 \neq 1$.

[**8.5**] If $x \in \mathbb{R}$, then, by A10, $0 \leqslant x$ or $x \leqslant 0$. Hence, either $x = 0$ or $x < 0$ or $0 < x$. In other words, $x \in \{0\} \cup \mathbb{R}^+ \cup \mathbb{R}^-$ and so $\mathbb{R} \subseteq \{0\} \cup \mathbb{R}^+ \cup \mathbb{R}^-$. Since $\{0\} \cup \mathbb{R}^+ \cup \mathbb{R}^-$ is a union of subsets of $\mathbb{R}$, $\mathbb{R} = \{0\} \cup \mathbb{R}^+ \cup \mathbb{R}^-$. If $x \in \mathbb{R}^+ \cap \mathbb{R}^-$, then $0 < x$ and $x < 0$. So, certainly $0 \leqslant x$ and $x \leqslant 0$. By A11, then, $x = 0$, which is impossible. So $\mathbb{R}^+ \cap \mathbb{R}^- = \emptyset$.

[**8.7**] By Theorem 8.6, $0 < 1$ and so $1 \in \mathbb{R}^+$. By A13, $0 + 1 \leqslant 1 + 1$, and so $1 \leqslant 2$. Since $0 \leqslant 1$ and $1 \leqslant 2$, by A12, $0 \leqslant 2$. If $0 = 2$, then $1 \leqslant 2$ entails $1 \leqslant 0$, which contradicts Theorem 8.6. Hence, $0 < 2$, and so $2 \in \mathbb{R}^+$.

[**8.10**] $x + (-1) \cdot x = x \cdot 1 + x \cdot (-1)$, using A6 and A7

$$= x \cdot (1 + (-1)), \text{ by A9}$$

$$= x \cdot 0, \text{ by A4}$$

$$= 0, \text{ by Theorem 8.1}$$

But $x + (-x) = 0$ by A4 and $(-x)$ is unique. Hence, $(-1) \cdot x = -x$.

[**8.11**] As an illustration, we prove that multiplication defines a closed binary operation on $\mathbb{N}$. We show that for each $m, n \in \mathbb{N}$, $m \cdot n \in \mathbb{N}$. For fixed $m \in \mathbb{N}$, let $T = \{n : n \in \mathbb{N} \wedge m \cdot n \in \mathbb{N}\}$. By A7, $m \cdot 1 = m \in \mathbb{N}$, and so $1 \in T$. If $k \in T$, then $m \cdot k \in \mathbb{N}$, and so

$$m \cdot (k + 1) = m \cdot k + m \cdot 1, \text{ by A9}$$

$$= m \cdot k + m, \text{ by A7}$$

Since $\mathbb{N}$ is closed under addition, $m \cdot k + m$ is an element of $\mathbb{N}$. Hence, $k + 1 \in T$ also. Thus, $\mathbb{N} \subseteq T$ and, since $T \subseteq \mathbb{N}$ by definition, $\mathbb{N} = T$. Because $m$ was arbitrary, we conclude that $\mathbb{N}$ is closed under multiplication.

[**8.12**] The supremum of a bounded set is unique, since, if there was a smaller supremum, then we would, in particular, have an upper bound smaller than the smallest upper bound. A similar argument shows that the infinum of a bounded set is unique.

[**8.13**] Given that any subset of $\mathbb{R}$ which is bounded above possesses a supremum, consider a set $S$ which is bounded below. Let $T = \{-x : x \in S\}$, the set of all negatives of elements of $S$. If $m$ is a lower bound for $S$, then $x \geqslant m$, $\forall x \in S$. Hence, $-x \leqslant -m$, $\forall -x \in T$. In other words, $T$ is bounded above. Thus, $T$ possesses a supremum, $\beta$, say. Since $-x \leqslant \beta$, $\forall -x \in T$, then $x \geqslant -\beta$, $\forall x \in S$. Hence, $-\beta$ is certainly a lower bound for $S$. There can be no larger lower bound, since its negative would be a smaller upper bound for $T$ than $T$'s supremum $\beta$, which is not the case. Hence, $-\beta$ is the infinum of $S$.

[**8.15**] As remarked at the end of Section 8.4, axioms A1, A2, A5, A6, A9–A13 hold for any non-empty subset of $\mathbb{R}$.

A3: $0 = 0 + 0\sqrt{5}$ is an element of $S$.
A4: If $a + b\sqrt{5} \in S$, then $(-a) + (-b)\sqrt{5} \in S$.
A7: $1 = 1 + 0\sqrt{5}$ is an element of $S$.
A8: If $a + b\sqrt{5} \in S$ is non-zero, then at least one of $a$ or $b$ is non-zero. Since $a, b \in \mathbb{Q}$, it is not possible for $a$ to equal $b\sqrt{5}$, and so $a - b\sqrt{5} \neq 0$. Now, the reciprocal of $a + b\sqrt{5}$ is

$$1/(a + b\sqrt{5}) = (a - b\sqrt{5})/(a - b\sqrt{5})(a + b\sqrt{5})$$
$$= (a - b\sqrt{5})/(a^2 - 5b^2)$$
$$= (a/(a^2 - 5b^2)) + ((-b)/(a^2 - 5b^2))\sqrt{5} \in S$$

A14: Follows, since $0 \in S$.

[**9.2**] The maps $x \mapsto x$ and $x \mapsto 4x$ are bijections from $E$ to a subset of $\mathbb{N}$ and from $\mathbb{N}$ to a subset of $E$, respectively.

[**9.5**] Just use induction. Let $A_1, A_2, A_3, \ldots$ be countable sets and let $S_k$ be the statement: '$A_1 \cup A_2 \cup \cdots \cup A_k$ is countable'. $S_1$ is true by hypothesis and we have proved that $S_2$ is true. Now $A_1 \cup A_2 \cup \cdots \cup A_{k+1} = (A_1 \cup A_2 \cup \cdots \cup A_k) \cup A_{k+1}$, so $S_k$ implies that this is the union of two countable sets, which is countable by $S_2$. Hence, $S_{k+1}$ is proven.

[**9.9**] With each non-zero integer $i$ associate a denumerable set of rationals $\{i/n : n \text{ is a natural number coprime with } i\}$. Each rational except zero is in one and only one of these denumerable sets. Hence, $\mathbb{Q} \setminus \{0\}$ has been expressed as a denumerable disjoint union of denumerable sets. By Theorem 9.3(b) this is denumerable, and by Theorem 9.2 this is not changed by throwing in zero. Alternatively, $\mathbb{Z}$ and $\mathbb{Z} \setminus \{0\}$ are denumerable and $\mathbb{Q}$ can be regarded as a subset of the Cartesian product of these two sets. (See Chapter 3.) Hence, $\mathbb{Q}$ is countable by Theorem 9.1. It is not finite, so must be denumerable.

[**9.11**] The integer $x$ appears in position $n$, where $n = 2|x| + \frac{1}{2}(1 - (-1)^x)$.

[**9.12**] If you propose $q$ as the 'next biggest' rational to $r$, I can always refute your claim with $\frac{1}{2}(q + r)$, which is also bigger than $r$ but less than $q$.

[**9.13**] The list consists of blocks of terms; each rational is expressed in its simplest form, $a/b$, with $b > 0$, and put in the $n$th block if $|a| + b = n$. Within each block the rationals are placed in pairs, $a/b$ followed by $-a/b$, and are written in order of increasing denominator. For example, the 10th block would be $\frac{1}{9}, -\frac{1}{9}, \frac{3}{7}, -\frac{3}{7}, \frac{7}{3}, -\frac{7}{3}, \frac{9}{1}, -\frac{9}{1}$. Clearly, every rational is represented in such a scheme and the coprime condition ensures that there are no repeated elements in the list.

[**9.14**] For $n = 2, 3, 4, \ldots, 1000$, $1000! + n$ is divisible by $n$.

[**9.15**] The sequence $(10^6 + 1)! + n : n = 2, 3, 4, \ldots, 10^6 + 1$ will do.

195

[9.17] $\log_{10} x \to \infty$ as $x \to \infty$, but $(x-1)/x \to 1$ as $x \to \infty$.

[9.18] If you divide the number by any one of the primes in $C$, there is necessarily a remainder of 1.

[9.19] The method of constructing the seventh row ensures that it differs from each of the other six in at least one place.

[9.20] Let $r$ be represented by decimals:

$$d = n.a_1 a_2 \ldots a_{n-1} b_n \ldots$$

and

$$d' = n.a_1 a_2 \ldots a_{n-1} a_n \ldots$$

with $b_n > a_n$. If $r$ has no finite expansion, there is some decimal place beyond the $n$th which is non-zero—say the $(n+k)$th place. Then

$$d > n.a_1 a_2 \ldots a_{n-1} b_n + 10^{-n-k}$$

and

$$d' \leqslant n.a_1 a_2 \ldots a_{n-1} b_n$$

which contradicts $r = d = d'$.

[9.21] $f$ is not one-to-one because, for example,

$$f(0.123\,191\,919\,19\ldots) = (0.139\,99\ldots, 0.211\,111\ldots) = (0.14, 0.211\,11\ldots)$$

and

$$f(0.124\,101\,010\,1\ldots) = (0.14, 0.211\,11\ldots)$$

It also fails to be *onto* $S$ because

$$f(0.090\,909\ldots) = (0, 0.999\ldots) = (0, 1)$$

which is not in $S$.

[9.22] Obtain the coordinates of $f(0.a_1 a_2 a_3 \ldots)$ by taking not alternate digits but alternate blocks of digits, defined as follows: block $1 = a_1$ unless $a_1 = 9$, in which case it is $a_1 a_2 \ldots a_n$, where $a_n$ is the next non-9 digit. Block 2 begins immediately after block 1 and is another single digit unless that digit is 9, in which case.... For example, the blocks of $0.903\,499\,987\,699\,402\,912\ldots$ are 90, 3, 4, 9998, 7, 6, 994, 0, 2, 91, 2, ....

[9.25] First show that $h$ is one-to-one. To do this, let $x, y \in A$ and suppose $h(x) = h(y)$. We have to prove that this implies $x = y$. Consider the three possibilities:

(i) $x, y \in \Omega$, and use the fact that $f$ is a bijection.
(ii) $x, y \in A \backslash \Omega$ and use the fact that $g$ is a bijection.
(iii) $x \in \Omega$, $y \in A \backslash \Omega$, and show that this can't arise if $h(x) = h(y)$.

Then we have to show that $h$ is onto $B$. Each element $b$ of $B$ is either in $f(\Omega)$ or $B \backslash f(\Omega)$. If the former, $b = f(x) = h(x)$ for some $x$ in $\Omega$; and if the latter, $b = g^{-1}(y) = h(y)$ for some $y$ in $A \backslash \Omega$.

[**9.26**] $S = \{\emptyset, \{2\}, \{1,3\}, \{2,3\}, \{1,2,3\}\}$.

[**9.27**] $\Omega = \{1, 2, 3\}$.

[**9.28**] Not including the 'add two' step would be equivalent to using the $d$'s instead of the $e$'s in the recipe for concocting our permutation of $\mathbb{N}$. We can use the same example as that given in the text to see what would go wrong in that case: $r = 0.31500704\ldots$, so that $d_1 = 3, d_2 = 1, d_3 = 5\ldots$, etc. The corresponding permutation would be

$$1\ 2\ 3\ 4\ 5\ 6\ 7\ 8\ 9\ 16\ 15\ 14\ 13\ 12\ 11\ 10\ 20\ 19\ 18\ 17\ \ldots$$

but this same permutation could equally well have come from the real number $0.40302704\ldots$ or from $0.010111104\ldots$, and from this you should see why it is essential to avoid 0's or 1's.

[**9.29**] $\mathscr{A}$ is a subset of $I$, so the identity map will do as our bijection from $\mathscr{A}$ to a subset of $I$. One way of getting a bijection going the other way is to first note that $\mathscr{A}$ consists of those reals in $(0, 1)$ which can only have 0 and 2 in their tricimals. Then, for each real number in $I$, take its decimal expansion (not containing a recurring 9, to ensure uniqueness)—say $0.d_1d_2d_3\ldots$; then replace this by the tricimal $0.00\ldots022\ldots200\ldots022\ldots 20\ldots$, which consists of alternating blocks of 0's and 2's, the length of block $i$ being $d_i$ except when $d_i = 0$, in which case the block length is 10. The resulting tricimal is clearly in $\mathscr{A}$, and from each such tricimal the decimal from which it was derived is uniquely recoverable. Hence, the process defines a bijection from $I$ to a subset of $\mathscr{A}$ and all we need to do now is invoke Schröder–Bernstein.

[**10.1**] We use strong induction (see Section 5.2.3) to prove that if $a_1 = a_2 = 1$ and $a_{n+2} = a_{n+1} + a_n$ for $n \geq 1$, then

$$a_n = \frac{1}{2^n\sqrt{5}}((1+\sqrt{5})^n - (1-\sqrt{5})^n)$$

Since

$$\frac{1}{2^1\sqrt{5}}((1+\sqrt{5})^1 - (1-\sqrt{5})^1) = \frac{2\sqrt{5}}{2\sqrt{5}} = 1$$

the result holds for $n = 1$. Now suppose that the result holds for all $n \leq k$ for some fixed $k \in \mathbb{N}$. Hence,

$a_{k+1} = a_k + a_{k-1}$

$$= \frac{1}{2^k\sqrt{5}}((1+\sqrt{5})^k - (1-\sqrt{5})^k) + \frac{1}{2^{k-1}\sqrt{5}}((1+\sqrt{5})^{k-1}$$
$$- (1-\sqrt{5})^{k-1})$$

$$= \frac{1}{2^k\sqrt{5}}((1+\sqrt{5})^k - (1-\sqrt{5})^k + 2(1+\sqrt{5})^{k-1} - 2(1-\sqrt{5})^{k-1})$$

$$= \frac{1}{2^k\sqrt{5}}((1+\sqrt{5})^{k-1}(3+\sqrt{5}) - (1-\sqrt{5})^{k-1}(3-\sqrt{5}))$$

$$= \frac{1}{2^k\sqrt{5}}\left((1+\sqrt{5})^{k+1}\left(\frac{3+\sqrt{5}}{(1+\sqrt{5})^2}\right) - (1-\sqrt{5})^{k+1}\left(\frac{3-\sqrt{5}}{(1-\sqrt{5})^2}\right)\right)$$

$$= \frac{1}{2^k\sqrt{5}}\left((1+\sqrt{5})^{k+1}\left(\frac{3+\sqrt{5}}{6+2\sqrt{5}}\right) - (1-\sqrt{5})^{k+1}\left(\frac{3-\sqrt{5}}{6-2\sqrt{5}}\right)\right)$$

Now,
$$\frac{(3\pm\sqrt{5})}{(6\pm 2\sqrt{5})} = \frac{1}{2}.$$

Hence,
$$a_{k+1} = \frac{1}{2^{k+1}\sqrt{5}}((1+\sqrt{5})^{k+1} - (1-\sqrt{5})^{k+1})$$

as required.

[10.3] $\sqrt{(n(n+1))} - n = \dfrac{(\sqrt{(n(n+1))} - n)(\sqrt{(n(n+1))} + n)}{(\sqrt{(n(n+1))} + n)}$

$$= \frac{(n(n+1) - n^2)}{(\sqrt{(n(n+1))} + n)}$$

$$= \frac{n}{(\sqrt{(n(n+1))} + n)}$$

$$= \frac{1}{(\sqrt{(1+1/n)} + 1)} \to \frac{1}{2}$$

[10.5] [$\varepsilon$ and $n$ are both positive.] So

$$2^{-n} < \varepsilon \Leftrightarrow 0 \leq \frac{1}{2^n} < \varepsilon$$

$$\Leftrightarrow \frac{1}{\varepsilon} < 2^n$$

$$\Leftrightarrow \log_2\left(\frac{1}{\varepsilon}\right) < n$$

[10.6] Given $\varepsilon > 0$, we can force $|1/n - 0| < \varepsilon$, provided that $1/n < \varepsilon$. So choose $N$ to be any integer greater than or equal to $1/\varepsilon$.

[10.7] If $a_n = 1$ for all $n \in \mathbb{N}$ and $\varepsilon > 0$, then $|a_n - 1| = 0 < \varepsilon$ for all $n \in \mathbb{N}$. So we can choose $N = 1$.

[10.9] $(a-b)^2 \geq 0$ for all $a, b \in \mathbb{R}$. Hence, $a^2 - 2ab + b^2 \geq 0$. In other words, $a^2 + b^2 \geq 2ab$.

[10.10] $a_n - L \leq y_n - L \leq b_n - L$, $\forall n \in \mathbb{N}$. Hence, $|y_n - L|$ cannot exceed the maximum of $|a_n - L|$ and $|b_n - L|$. So, for any $\varepsilon > 0, \exists N$ such that $|a_N - L| < \varepsilon$ and $|b_N - L| < \varepsilon$. But then $n > N \Rightarrow |y_n - L| < \varepsilon$.

[**10.11**] $|a+b|^2 = (a+b)^2 = a^2 + 2ab + b^2$. Now, $ab \leq |ab|$, with strict inequality only when $a$ and $b$ have opposite signs. Hence,

$$|a+b|^2 \leq a^2 + 2|a||b| + b^2 = |a|^2 + 2|a||b| + |b|^2 = (|a|+|b|)^2$$

and so $|a+b| \leq |a| + |b|$. Also,

$$|a-b| = |a+(-b)| \leq |a| + |-b| = |a| + |b|$$

[**10.13**] The $n$th partial sum of $\sum_{k=0}^{\infty} x^k$ is $s_n = 1 + x + x^2 + \cdots + x^{n-1}$. Hence, $xs_n = x + x^2 + \cdots + x^n$. Subtraction yields that $(1-x)s_n = 1 - x^n$. If $x \neq 1$, then $s_n = (1-x^n)/(1-x)$. Since $\lim_{n \to \infty} x^n$ exists if and only if $|x| < 1$, $\sum_{k=0}^{\infty} x^k$ is convergent for $|x| < 1$ (with sum $1/(1-x)$) and divergent for $x > 1$ and $x \leq -1$.

If $x = 1$, then $s_n = n$ and so $\sum_{k=0}^{\infty} x^k$ is divergent for $x = 1$.

[**10.14**] From exercise [**10.13**], $\sum_{k=0}^{\infty} x^k$ is convergent for $|x| < 1$ and its sum for such $x$ is $1/(1-x)$. Since $\sum_{k=0}^{\infty} 2^k$ is divergent, its 'sum' cannot be $1/(1-2)$, as claimed in (c). Similarly, $\sum_{k=0}^{\infty}(-1)^k$ is divergent, giving the contradiction in the last part of (b).

[**10.15**] Let $s_k, m_k, p_k, n_k$ be the $k$th partial sums of the series $\sum_{n=1}^{\infty} a_n$, $\sum_{n=1}^{\infty} |a_n|, \sum_{n=1}^{\infty} a_n^+, \sum_{n=1}^{\infty} a_n^-$, respectively, where $a_n^+ = a_n$ if $a_n \geq 0$, and zero otherwise, and $a_n^- = a_n$ if $a_n \leq 0$, and zero otherwise. In other words, $\sum_{n=1}^{\infty} a_n^+$ is the series derived from $\sum_{n=1}^{\infty} a_n$ by replacing all the negative terms by zero, and similarly for $\sum_{n=1}^{\infty} a_n^-$.

Now, $\sum_{n=1}^{\infty} a_n$ absolutely convergent means that $\sum_{n=1}^{\infty} |a_n|$ converges and so $(m_k)$, a monotonic increasing sequence, must be bounded. Now, $(p_k)$ is a monotonic increasing sequence and $(n_k)$ is a monotonic decreasing sequence. Also $m_k = p_k - n_k \geq p_k$ and $s_k = p_k + n_k$. If $(p_k)$ diverged (necessarily to $+\infty$), then, because $m_k \geq p_k$, $(m_k)$ would also diverge. This contradicts the absolute convergence of $\sum_{n=1}^{\infty} a_n$. The same contradiction is easily deduced from the divergence of $(n_k)$, so, if $\sum_{n=1}^{\infty} a_n$ is absolutely convergent, then both $\sum_{n=1}^{\infty} a_n^+$ and $\sum_{n=1}^{\infty} a_n^-$ must converge. [Note that we also get from this proof the important fact that absolute convergence implies convergence, since $(m_k)$ convergent gives $(p_k)$ and $(n_k)$ convergent and, hence, $(s_k)$ converges.]

[**10.16**] To prove this, simply note that the convergence of both $(p_k)$ and $(n_k)$ implies the convergence of $(m_k)$, since $m_k = p_k - n_k$. But the definition of conditional convergence says that $(m_k)$ diverges.

[11.1]

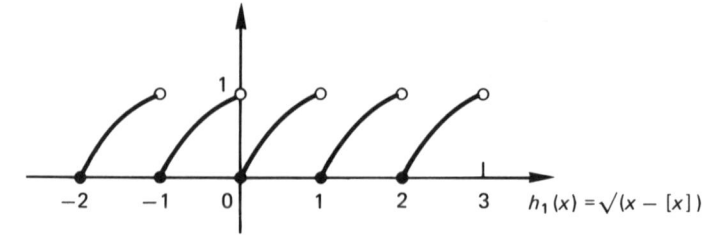

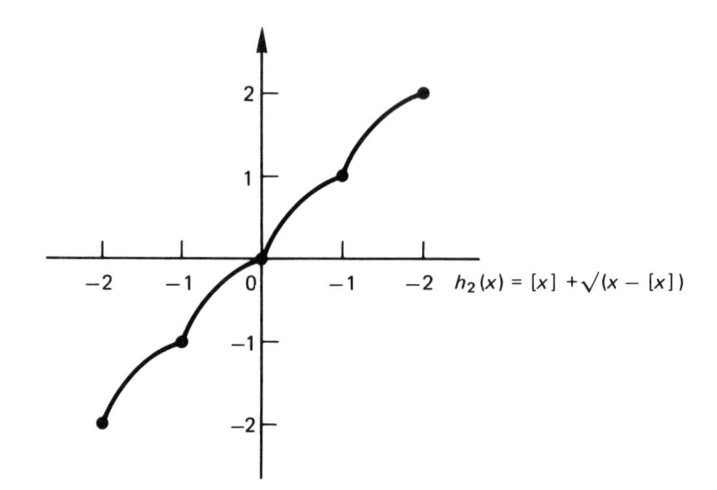

$h_1(x) = \sqrt{(x - [x])}$

$h_2(x) = [x] + \sqrt{(x - [x])}$

[11.3] $\dfrac{\sin x}{x} = 0 \Leftrightarrow x = n\pi,\ n \in \mathbb{Z},\ n \neq 0.$

[11.4] $\left|\dfrac{\sin x}{x}\right| \leq \dfrac{1}{|x|}$, and so $\dfrac{-1}{|x|} \leq \dfrac{\sin x}{x} \leq \dfrac{1}{|x|}$, for $x \neq 0.$

[11.5]

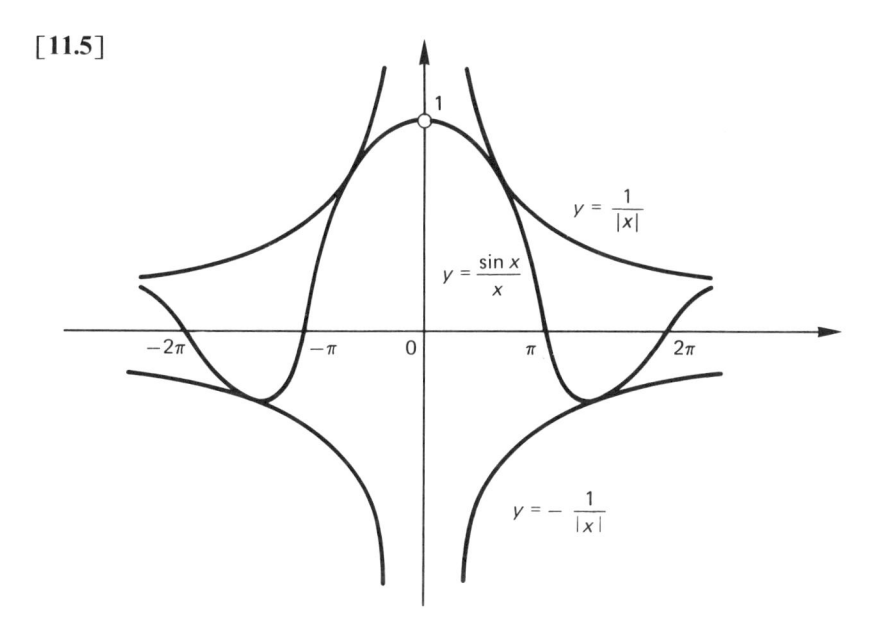

[11.6] The decimal expansion of an irrational is infinite.

[11.7] $D(x) = 1/q$ at precisely $(q-1)$ points in the interval $(0, 1)$ if and only if $q$ is a prime number. Only then are the rationals, $1/q, 2/q, \ldots, (q-1)/q$ all in their lowest form.

[11.8] If $f(x) = mx + c$, then $f(x_0) = mx_0 + c$. Now,

$$|f(x) - f(x_0)| < \varepsilon \Leftrightarrow |mx + c - mx_0 - c| < \varepsilon$$
$$\Leftrightarrow |m||x - x_0| < \varepsilon$$

So choose $\delta = \varepsilon/|m|$.

[11.9] Since $x \mapsto x$, the identity function, and $x \mapsto k$, a constant function, are everywhere continuous, then repeated use of the product rule gives that $x \mapsto kx^r$ is continuous for all $k \in \mathbb{R}$ and for all $r \in \mathbb{N}$. Repeated application of the sum rule gives that all polynomials are continuous.

[11.10] If $n \leqslant 1/2\pi\delta$ for all $n \in \mathbb{N}$, then $\mathbb{N}$ would be bounded above!

[11.12] If $1/q \geqslant \varepsilon$ for all $q \in \mathbb{N}$, then $q \leqslant 1/\varepsilon$ for all $q \in \mathbb{N}$, which contradicts the fact that $\mathbb{N}$ is not bounded above.

# INDEX

Abelian group   99
Alice   §1, §5, §7, §9, 107
Archimedean property   109
Arithmetic on intervals   92
Associative   11, 65, 69, 85
Axioms   12, 94, 95, 107
   completeness   96, 108, 146, 169
   for $\mathbb{R}$   §8, 99, 102

Bijection   74, 80, 113
Bolzano, Bernhard   130, 161
Bounded above   108
Bounded below   108
Bounded set   41, 49
Brouwer, L. E. J.   130

Cantor, Georg   111
Cardinal number   115
Cartesian product   13, 27
Cesaro curve   175
Closed operation   27, 98, 105
Commensurable   39
Commutative   11, 69, 85, 91
Complement   11
Complex numbers   9, 19
Composite number   16
Conjecture   7, 69
Connectives
   and   2, 67
   if and only if   5, 67
   implication   3, 4, 67
   not   2, 67
   or   2, 67
Continuity   §11, 168, 169
   of Cantor's bijection   130
   and dimension   130
   of graphs   162
   of number line   50
   of Peano's curve   130

Contradictory   7
Contrapositive   5, 56
Converse   5, 34
Coprime   56
Counterexample   7

De Morgan laws   11
Decimal   132
Dedekind cuts   50
Dedekind, Richard   129
Density
   of primes   125
   of $\mathbb{Q}$   36
Diagonal method   127
Difference of sets   13
Dimension
   fractal   174
   topological   130, 173
Dirichlet's function   167, 171
Disjoint
   cycles   83
   pairwise   121
   sets   22
Distributive   11
Distributive law   152
Dual   12
Duality, principle of   12

Equivalence
   class   33, 90, 115
   of conjugate permutations   86
   logical   3, 69
   of nests   47, 90
   relation   27, 33, 86, 91, 115
   topological   130
Euclid's algorithm   44
Euclid's infinity of primes   126
Euler, Leonard   7
Euler number   160

# INDEX

Factorisation §2, 75
  unique §2, 37, 60, 76, 82
Family 15
Fermat, Pierre de 7
Fermat prime 7
Fermat's last theorem 7
Fermat's little theorem 7
Fibonacci, Leonardo 142
Fibonacci numbers 63
Fibonacci, and rabbit breeding 52
Fibonacci sequence 53, 142, 149
Field 99, 102
  complete 108
  ordered 102
Fractals 173
  dimension 174
  generator 173
  initiator 173
Fundamental Theorem of Arithmetic §2

Gauss, Carl Friedrich 15
Golden ratio 142, 150
Golden rectangle 150
Googol 144
Googolplex 144
Graph 66, 70
  circuit 70
  component 66
  continuity of §11, 163, 168
  Dirac's theorem 70
  edge 66
  Hamiltonian 70
  vertex 66
Greatest common measure 39

Highest common factor 16, 44

Idempotent 11
Identity
  permutation 75
  rule 11
Induction §5, 8, 105, 122, 127, 142
  backwards 63
  basis of 55
  hypothesis 55
  justification for 60
  step 55
  strong 61
  weak 61
Inductive definition 68
Inequality 103
Infinitesimal 109
Infinity 109, §9

Infinum 108, 146
Integers 9, 105
Intermediate value theorem 161
Intersection 10, 91
Interval theorem 168
Intervals
  closed 45
  half-open 131
  infinite 49
  nested 45
  open 45
  rational 90, 92
Irrational numbers 37, 50, 130

Koch snowflake curve 173

Leibniz, Gottfried 130
Leonardo of Pisa 52
Linear 25
Logic §1, 67
Logical formula 67
Lower bound 108
Lüroth 130

Mandelbrot, Benoit 173
Mardesic 130
Multiplication, a model for 23

Natural numbers 5, 9, 14, 105
Negative numbers 23
Nest 46, §7, 122
  equivalent 90
  rational 90
Number
  discovery of 47
  invention of 47
  line 22, 50

One-to-one
  correspondence 74, 113
  function 129
Operations
  on $\mathbb{Q}$ 28
  on $\mathbb{Z}$ 27
Ordered pairs 27
Ordinal 22

Parabola 163
Partitions 31
Peano curve 175
Peano, Giuseppe 130
Peano space 130
Periodic function 163

Permutations §6
  composition 74
  conjugate 86
  cycle 82
  even 78
  as a function 73
  identity 76
  image 73
  inverse 84
  odd 78
  order 85
  parity 78
  transposition 76
Polya, George 63
Predicate 6, 8
Prime 7, 16, 76, 124
Proof 2
  constructive 83
  contradiction 5, 37, 90, 97, 109
  direct 6, 55
  inductive §5
  *reductio ad absurdum* 5
Propositional calculus 67
Pythagoras' theorem 4, 37

Quantifier
  existential 6
  universal 6

Rational numbers 9, 36, 105
Real numbers §4, 9, §8
Relation 32
  equivalence 27, 33, 86
  order 95, 102
  reflexive 32, 91
  symmetric 32, 91
  transitive 32, 91
Richardson, Lewis 173

Schröder–Bernstein theorem 132
Sequences 141, 169
  bounded 49, 146
  Cauchy 147
  convergent 142, 144, 145
  Fibonacci 53, 63, 142, 149
  limit of 49, 145, 169
  Lucas 61
  monotonic 49, 131, 146
  subsequence 148
Series 151, 152
  absolutely convergent 155
  conditionally convergent 155
  convergent 152
  divergent 153

Fourier 156
  geometric 153
  harmonic 157
  $p$- 157
  partial sum 152
  power 154
  rearrangement 155
  sum 152
  Taylor–Maclaurin 156
Sets 8
  algebra of 11
  bounded 41, 49, 96
  countable 116
  denumerable 116, 121
  disjoint 22
  empty 12
  equinumerous 113
  infinite 14, §9, 35
  notation 8
  operation 9
  pairwise disjoint 121
  power 128
  subset 9, 105
  universal 9
Sierpinski 130
Smallest upper bound 96
Statement
  atomic 67
  compound 2
  simple 2
Supremum 96, 108, 146
Symmetric difference 13

Tautology 69
Theorem 5, 98
Triangle inequality 149
Trichotomy law 132
Tricimal 139
Truth tables 2
Truth values 2
Tweedledee §1, §5, §7, §9, 107
Tweedledum §1, §5, §7, §9, 107

Union 10
Upper bound 96, 146

Venn diagram 9
Venn, John 9

Weierstrass, Karl 129
Well-defined operation 27, 28, 78, 93, 94
Whitehead, Alfred 130

Zeno's paradox 153
Zermelo, Ernst 132